VOLUME SIXTY NINE

ADVANCES IN
FOOD AND NUTRITION
RESEARCH

VOLUME SIXTY NINE

ADVANCES IN
FOOD AND NUTRITION
RESEARCH

Edited by

JEYAKUMAR HENRY

Singapore Institute for Clinical Sciences, Singapore
Oxford Brookes University, UK

AMSTERDAM • BOSTON • HEIDELBERG • LONDON
NEW YORK • OXFORD • PARIS • SAN DIEGO
SAN FRANCISCO • SINGAPORE • SYDNEY • TOKYO
Academic Press is an imprint of Elsevier

Academic Press is an imprint of Elsevier
225 Wyman Street, Waltham, MA 02451, USA
525 B Street, Suite 1800, San Diego, CA 92101-4495, USA
32 Jamestown Road, London, NW1 7BY, UK
The Boulevard, Langford Lane, Kidlington, Oxford, OX5 1GB, UK
Radarweg 29, PO Box 211, 1000 AE Amsterdam, The Netherlands

First edition 2013

ISBN: 978-0-12-410540-9
ISSN: 1043-4526

For information on all Academic Press publications
visit our website at store.elsevier.com

Transferred to Digital Printing in 2013

CONTENTS

CONTRIBUTORS

R.M. Anjana
Madras Diabetes Research Foundation, WHO Collaborating Centre for Non-communicable Diseases Prevention and Control, IDF Centre of Education, Gopalapuram, Chennai, India

A. Barkouti
UMR 1208 IATE, Université Montpellier 2, Montpellier, and AgroParisTech, UMR GENIAL, Massy, France

B. Cuq
Montpellier SupAgro, UMR 1208 IATE; INRA, UMR 1208 IATE, and UMR 1208 IATE, Université Montpellier 2, Montpellier, France

M. Delalonde
UMR QualiSud, Université Montpellier 1, Montpellier, France

G. Delaplace
INRA, UR 638 PIHM, Villeneuve d'Ascq, France

Nikhil V. Dhurandhar
Pennington Biomedical Research Center, Louisiana State University System, Baton Rouge, Louisiana, USA

E. Dumoulin
AgroParisTech, UMR GENIAL, Massy, France

C. Gaiani
LiBio, Université de Lorraine, Nancy, France

L. Galet
Ecole des Mines Albi, UMR RAPSODEE, Albi, France

Frank L. Greenway
Pennington Biomedical Research Center, Louisiana State University System, Baton Rouge, Louisiana, USA

Apoorva Gupta
Food Engineering and Technology Department, Institute of Chemical Technology, Matunga, Mumbai, India

R. Jeantet
AgroCampusOuest, and INRA, UMR 1253 STLO, Rennes, France

Lalit D. Kagliwal
Food Engineering and Technology Department, Institute of Chemical Technology, Matunga, Mumbai, India

K. Krishnaswamy
Madras Diabetes Research Foundation, WHO Collaborating Centre for
Non-communicable Diseases Prevention and Control, IDF Centre of Education,
Gopalapuram, Chennai, India

Ann G. Liu
Pennington Biomedical Research Center, Louisiana State University System, Baton Rouge,
Louisiana, USA

N.G. Malleshi
Madras Diabetes Research Foundation, WHO Collaborating Centre for
Non-communicable Diseases Prevention and Control, IDF Centre of Education,
Gopalapuram, Chennai, India

S. Mandato
Montpellier SupAgro, UMR 1208 IATE; INRA, UMR 1208 IATE, and UMR 1208 IATE,
Université Montpellier 2, Montpellier, France

V. Mohan
Madras Diabetes Research Foundation, WHO Collaborating Centre for
Non-communicable Diseases Prevention and Control, IDF Centre of Education,
Gopalapuram, Chennai, India

I. Murrieta-Pazos
LiBio, Université de Lorraine, Nancy, and Ecole des Mines Albi, UMR RAPSODEE,
Albi, France

L. Palaniappan
Palo Alto Medical Foundation Research Institute, Palo Alto, CA, USA

J. Petit
LiBio, Université de Lorraine, Nancy; AgroCampusOuest, UMR 1253 STLO, Rennes;
INRA, UMR 1253 STLO, Rennes, and INRA, UR 638 PIHM, Villeneuve d'Ascq, France

Candida J. Rebello
Pennington Biomedical Research Center, Louisiana State University System, Baton Rouge,
Louisiana, USA

E. Rondet
UMR QualiSud, Université Montpellier 1, Montpellier, France

T. Ruiz
Montpellier SupAgro, UMR 1208 IATE; INRA, UMR 1208 IATE, and UMR 1208 IATE,
Université Montpellier 2, Montpellier, France

J. Scher
LiBio, Université de Lorraine, Nancy, France

P. Schuck
AgroCampusOuest, and INRA, UMR 1253 STLO, Rennes, France

S. Shobana
Madras Diabetes Research Foundation, WHO Collaborating Centre for
Non-communicable Diseases Prevention and Control, IDF Centre of Education,
Gopalapuram, Chennai, India

Rekha S. Singhal
Food Engineering and Technology Department, Institute of Chemical Technology, Matunga, Mumbai, India

V. Sudha
Madras Diabetes Research Foundation, WHO Collaborating Centre for Non-communicable Diseases Prevention and Control, IDF Centre of Education, Gopalapuram, Chennai, India

C. Turchiuli
AgroParisTech, UMR GENIAL, Massy, France

PREFACE

In 1912, the Polish scientist Casimir Funk coined the term "vitamine" after "vita" meaning life and "amine" from the first isolated compound, thiamine from rice bran. Subsequently, it was recognized that all vitamins were not amines and a range of vitamins of varying composition were isolated and characterized. It is now believed that the origins of modern nutrition can be traced to the discovery of vitamins. In the intervening century, modern nutrition has been transformed from therapeutic to preventive nutrition. Mankind has always been interested in food. The science of nutrition is the convergence of two interests in our lives—diet and health. The current emphasis on preventive nutrition has attracted considerable interest from a range of disciplines including clinicians, food technologists, and physiologists. We have witnessed a close synergy between food technology and human nutrition. Modern food technology has enabled us to enjoy a variety of foods with exceptional nutritional and culinary attributes. In addition, it has enabled the fabrication of new foods and the introduction of "functional foods." The concept of functional foods is often ascribed as a newly evolving discipline. However, both Indian and Chinese traditional medicines have articulated that "medicine and food has a common origin." In keeping with its long tradition, *Advances in Food and Nutrition Research* will maintain its role in providing topical, relevant, and cutting-edge science in the interface in food and nutrition. Readers of this monograph are encouraged to submit innovative chapters wherever they may be. Science is a global activity. If we are to advance the science of nutrition and food, it is imperative that we nurture collaboration between nations and communities. Only then can we reap its benefits.

C.J. HENRY
Singapore, Oxford

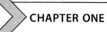

CHAPTER ONE

Finger Millet (*Ragi, Eleusine coracana* L.): A Review of Its Nutritional Properties, Processing, and Plausible Health Benefits

S. Shobana[*], K. Krishnaswamy[*,1], V. Sudha[*], N.G. Malleshi[*], R.M. Anjana[*], L. Palaniappan[†], V. Mohan[*]

[*]Madras Diabetes Research Foundation, WHO Collaborating Centre for Non-communicable Diseases Prevention and Control, IDF Centre of Education, Gopalapuram, Chennai, India
[†]Palo Alto Medical Foundation Research Institute, Palo Alto, CA, USA
[1]Corresponding author: e-mail address: sri21kk@yahoo.com

Contents

Advances in Food and Nutrition Research, Volume 69
ISSN 1043-4526
http://dx.doi.org/10.1016/B978-0-12-410540-9.00001-6

Abstract

Finger millet or *ragi* is one of the ancient millets in India (2300 BC), and this review focuses on its antiquity, consumption, nutrient composition, processing, and health benefits. Of all the cereals and millets, finger millet has the highest amount of calcium (344 mg%) and potassium (408 mg%). It has higher dietary fiber, minerals, and sulfur containing amino acids compared to white rice, the current major staple in India. Despite finger millet's rich nutrient profile, recent studies indicate lower consumption of millets in general by urban Indians. Finger millet is processed by milling, malting, fermentation, popping, and decortication. Noodles, vermicilli, pasta, Indian sweet (*halwa*) mixes, *papads*, soups, and bakery products from finger millet are also emerging. *In vitro* and *in vivo* (animal) studies indicated the blood glucose lowering, cholesterol lowering, antiulcerative, wound healing properties, etc., of finger millet. However, appropriate intervention or randomized clinical trials are lacking on these health effects. Glycemic index (GI) studies on finger millet preparations indicate low to high values, but most of the studies were conducted with outdated methodology. Hence, appropriate GI testing of finger millet preparations and short- and long-term human intervention trials may be helpful to establish evidence-based health benefits.

1. INTRODUCTION

Millets are minor cereals of the grass family, Poaceae. They are small seeded, annual cereal grasses, many of which are adapted to tropical and arid climates and are characterized by their ability to survive in less fertile soil (Hulse, Laing, & Pearson, 1980). Millets include sorghum (*Jowar*), pearl millet (*Bajra*), finger millet (*Ragi*), foxtail millet (*Kakum*), proso millet (*Chena*), little millet (*Kutki*), kodo millet (*Kodon*), barnyard millet (*Sanwa*), and brown top millet (Gopalan, Rama Sastri, & Balasubramanian, 2009; Hulse et al., 1980; http://www.swaraj.org/shikshantar/millets.pdf; Accessed on 27.03.12). Ragi or finger millet (*Eleusine coracana* L.) is one of the common millets in several regions of India. It is also commonly known as *Koracan* in Srilanka and by different names in Africa and has traditionally been an important millet staple food in the parts of eastern and central Africa and India (FAO, 1995). Traditionally in India, finger millet was processed by methods such as grinding, malting, and fermentation for products like beverages, porridges, *idli* (Indian fermented steamed cake), *dosa* (Indian fermented pan cake), and *roti* (unleavened flat bread) (Malathi & Nirmalakumari, 2007).

Research evidence suggests that whole grains and cereal fiber consumption are inversely associated with BMI, waist circumference, total cholesterol, metabolic syndrome, mortality from cardiovascular diseases, insulin resistance, and incidence of type 2 diabetes (de Munter, Hu, Spiegelman, Franz, & van Dam,

2007; Newby et al., 2007; Pereira et al., 2002; Sahyoun, Jacques, Zhang, Juan, & McKeown, 2006). Whole-grain cereals such as brown rice and millets are nutritionally superior to the widely consumed polished white rice. Studies have also shown that the dietary glycemic load (a measure of both quality and quantity of carbohydrates) and the higher intake of refined grains such as white rice were associated with the risk of type 2 diabetes and metabolic syndrome among urban south Asian Indians (Mohan et al., 2009; Radhika, Van Dam, Sudha, Ganesan, & Mohan, 2009). Several studies have reported on the beneficial role of low glycemic index (GI—a measure of carbohydrate quality) foods/diets in the nutritional management of diabetes and several other chronic diseases. The rate of glucose absorption is usually decreased by low GI foods and hence reduced insulin demand (Augustin, Franceschi, Jenkins, Kendall, & Vecchia, 2002). A calorie-restricted diet, moderately lower in carbohydrate, was also found to be helpful in decreasing insulin resistance and other metabolic abnormalities in overweight South Asian Indians (Backes et al., 2008). Dixit, Azar, Gardner, and Palaniappan (2011) have given practical strategies to reduce the burden of chronic disease by the incorporation of whole, ancient grains such as sorghum and millets like finger millet into the modern Asian Indian diets. Kannan (2010) has discussed the various nutritional and health benefits of finger millet with special reference to chronic disease preventive potential.

Unfortunately, many of the traditional Indian grains including millets have not been evaluated for GI using appropriate methods. In addition, development and availability of low GI foods are limited especially from millets in India. Finger millet being a low cost millet with higher dietary fiber contents, several micronutrients and phytonutrients with practically no reports of its adverse effect, deserves attention. This review attempts to explore the plausible health benefits of processed finger millet with particular reference to its nutritional and glycemic properties. Literature search was conducted in Pubmed, Science Direct data bases, and Google searches using suitable keywords. Studies published from 1939 to 2012 are included in this review.

2. HISTORY OF FINGER MILLET

Finger millet, one of the oldest crops in India is referred as "*nrtta-kondaka*" in the ancient Indian Sanskrit literature, which means "Dancing grain," was also addressed as "*rajika*" or "*markataka*" (Achaya, 2009). Earliest report of finger millet comes from Hallur in Karnataka of India dating approximately 2300 BC (Singh, 2008). There is some debate as to the origin of finger millet, and there are theories that finger millet might have traveled to India by sea from Arabia or South Africa or across the Indian Ocean in both directions (Achaya, 2009). Fuller (2002, 2003) has extensively reviewed the archeological work carried out in India and elsewhere on the origin of *Eleusine,*and reports its African origin and provides linguistic evidence for the term "ragi" from the root source term-dègi for finger millet in a number of Bantu languages from southern Tanzania and northern Malawi and its other variants in Indian subcontinent. Fuller is skeptical about most of the findings on finger millet and reported that the grains isolated from Hallur (1000 BC), Malhar (800 BC–1600 BC), and Hulaskhera (700 BC) in India to be authentic (Fuller, 2003). Although *Ragi* (finger millet), *jowar* (sorghum), and *bajra* (pearl millet) are of African origin, they have long been domesticated in India (Achaya, 2009).

Finger millet was a well-domesticated plant in various states of India and popularly called as *nachni* (meaning dancer) in the state of Maharastra, "*umi*" in Bihar, etc. The grains were gently roasted (sometimes after it was sprouted and dried), ground, sieved. The pinkish flour (from red finger millet) was eaten as a ball or gruel, either sweetened or salted. Finger millet was also popular as weaning foods (Achaya, 2009). The ancient Tamil literature from India, "*Kuruntogai*," addresses red finger millet as "*Kelvaragu*". *Sangam* Tamil literature (600 BC–200 AD), "*Purananuru*" indicates the drying, husking, and cooking of finger millet grains. In ancient India, finger millet cooked in milk was served with honey to poets (Achaya, 1992). It was then and now being used in Karnataka.

3. MILLET CONSUMPTION IN INDIA

Reports of National Nutrition Monitoring Bureau (NNMB, 2006) indicated that consumption of millets in general was higher in the states of Gujarat (maize, pearl millet), Maharastra (sorghum), Karnataka (finger millet) but almost nil in the states of Kerala, West Bengal, Orissa, and Tamil Nadu where rice forms the major staple. The consumption of millets in Gujarat and Maharashtra (200 and 132 g/CU (consumption unit is a coefficient and 1 CU corresponds to energy requirement of 2400 kcal/day of an Indian male doing sedentary work) was higher compared to that of Karnataka (75 g/CU/day), Madhya Pradesh (32 g/CU/day), and Andhra Pradesh (16 g/CU/day). Tamil Nadu (3 g/CU/day) and Orissa (1 g/CU/day) showed negligible amounts of consumption.

Though Indians continue to consume cereals as the main staple providing 70–80% of total energy intake in majority of Indian diets

(Gopalan et al., 2009), the consumption of millets is very low compared to rice and as evident by our recent study on dietary profile of urban Indians (from the Chennai Urban Rural Epidemiology Study (CURES)) which revealed that, the millets contributed to only about 2% of total calories (6.7 g/d) (Radhika et al., 2011), while almost half of the daily calories were derived from refined grains such as polished white rice (253.4 g/day) (Radhika et al., 2009).

4. NUTRITIONAL SIGNIFICANCE OF STRUCTURAL FEATURES OF FINGER MILLET

The seed coat, embryo (germ), and the endosperm are the main botanical components of the millet kernel. Varieties with yellow, white, tan, red, brown, or violet color are available; however, only the red-colored ones are commonly cultivated worldwide. The pericarp (the outer most covering of the millet) is of little nutritional significance. The seed coat or the testa is multilayered (five layered), which is unique compared to other millets such as sorghum, pearl millet, proso millet, and foxtail millet (FAO, 1995) and may this could be one of the possible reasons for the higher dietary fiber content in finger millet. The seed coat is tightly bound to the aleurone layer (a layer between the seed coat and endosperm) and the starchy endosperm, which is further divided into corneous and floury regions. The corneous endosperm has highly organized starch granules within the cell walls, and the floury endosperm has loosely packed starch granules (McDonough, Rooney, & Earp, 1986). The sizes of the finger millet starch granule in different regions of the kernel greatly vary compared to pearl and proso millets and ranges from 3 to 21 μm (Serna-Saldiver, McDonough, & Rooney, 1994). The starch granules in the floury endosperm of millets in general are bigger compared to the ones present in the corneous endosperm and hence more susceptible to enzymatic digestion (FAO, 1995). However, further research is required to study the enzyme (digestive enzymes) susceptibility characteristics of starch present in the corneous and floury endosperm regions of finger millet individually. Generally, finger millet is milled with the seed coat (rich in dietary fiber and micronutrients) to prepare flour and the whole meal is utilized in the preparation of foods. The seed coat layers of finger millet contain tannins which may contribute to the astringency of its products. Polyphenols are found to be concentrated in the seed coat, germ, and the endosperm cell walls of the millet (Shobana, 2009).

5. NUTRIENT COMPOSITION OF FINGER MILLET

The detailed nutrient composition of finger millet compared to other cereals and millets is summarized in Table 1.1.

5.1. Carbohydrates

Finger millet is a rich source of carbohydrates and comprises of free sugars (1.04%), starch (65.5%), and non-starchy polysaccharides (Malleshi, Desikachar, & Tharanathan, 1986) or dietary fiber (11.5%) (Gopalan et al., 2009). Wankhede, Shehnaj, and Raghavendra Rao (1979a) studied the carbohydrate profile of a few varieties of finger millet and reported 59.5–61.2% starch, 6.2–7.2% pentosans, 1.4–1.8% cellulose, and 0.04–0.6% lignins. The dietary fiber content of finger millet (11.5%) is much higher than the fiber content of brown rice, polished rice, and all other millets such as foxtail, little, kodo, and barnyard millet. However, the dietary fiber content of finger millet is comparable to that of pearl millet and wheat (Table 1.1). The carbohydrate content of finger millet is comparable to that of wheat but lower than that of polished rice (Table 1.1). Finger millet starch comprises amylose and amylopectin. The amylose content of finger millet starch is lower (16%) (Wankhede, Shehnaj, & Raghavendra Rao, 1979b) compared to the other millets such as sorghum (24.0%), pearl millet (21.0%), proso millet (28.2%), foxtail millet (17.5%), and kodo millet (24.0%) (FAO, 1995). Finger millet (Purna variety) starch had the highest set back viscosity (560 BU) during cooling (from 930 to 500 °C) which is suggestive of its tendency to retrograde (Wankhede et al., 1979b) (retrogradation of starch is known to induce resistance starch formation).

5.2. Proteins

Varietal variation exists in protein content of finger millet. Prolamins are the major fractions of finger millet protein (Virupaksha, Ramachandra, & Nagaraju, 1975). In general, cereals and millets contain lower amount of lysine compared to legumes and animal protein (ICMR, 2010). Albumin and globulin fractions contain several essential amino acids, while the prolamin fraction contains higher proportion of glutamic acid, proline, valine, isoleucine, leucine, and phenylalanine but low lysine, arginine, and glycine. The chemical score (a measure of protein quality, calculated as the ratio of amount of amino acid in a test protein over reference protein expressed as

Table 1.1 Nutrients, vitamins, minerals, and essential amino acid composition of finger millet (all values are per 100 g of edible portion)

Parameter	*Ragi*/finger millet	Proso millet	Foxtail millet	Little millet	Kodo millet	Barnyard millet	Pearl millet	Rice (raw) milled	Wheat
Proximates									
Moisture (g)	13.1	11.9	11.2	11.5	12.8	11.9	12.4	13.7	12.8
Protein (N × 6.25) (g)	7.3	12.5	12.3	7.7	8.3	6.2	11.6	6.8	11.8
Fat (g)	1.3	1.1	4.3	4.7	1.4	2.2	5.0	0.5	1.5
Minerals (g)	2.7	1.9	3.3	1.5	2.6	4.4	2.3	0.6	1.5
Dietary fiber (g)	11.5	–	2.4	2.53	2.47	1.98	11.3	4.1	12.5
Carbohydrates (g)	72.0	70.4	60.9	67	65.9	65.5	67.5	78.2	71.2
Energy (kcal)	328	341	331	341	309	307	361	345	346
Vitamins									
Carotene (μg)	42	0	32	0	0	0	132	0	64
Thiamine (mg)	0.42	0.20	0.59	0.30	0.33	(0.33)	0.33	0.06	0.45
Riboflavin (mg)	0.19	0.18	0.11	0.09	0.09	(0.10)	0.25	0.06	0.17
Niacin (mg)	1.1	2.3	3.2	3.2	2.0	4.2	2.3	1.9	5.5
Total B6 (mg)	–	–	–	–	–	–	–	–	(0.57)
Folic acid (μg)									

Continued

Table 1.1 Nutrients, vitamins, minerals, and essential amino acid composition of finger millet (all values are per 100 g of edible portion)—cont'd

Parameter	Ragi/finger millet	Proso millet	Foxtail millet	Little millet	Kodo millet	Barnyard millet	Pearl millet	Rice (raw) milled	Wheat
Free	5.2	–	4.2	2.2	7.4	–	14.7	4.1	142
Total	18.3	–	15.0	9.0	23.1	–	45.5	8.0	36.6
Vitamin C (mg)	0	0	0	0	0	0	0	0	0
Choline (mg)	–	748	–	–	–	–	–	–	–
Minerals and trace elements									
Calcium (mg)	344	14	31	17	27	20	42	10	41
Phosphorus (mg)	283	206	290	220	188	280	296	160	306
Iron (mg)	3.9	0.8	2.8	9.3	0.5	5.0	8.0	0.7	5.3
Magnesium (mg)	137	153	81	133	147	82	137	64	138
Sodium (mg)	11	8.2	4.6	8.1	4.6	–	10.9	–	17.1
Potassium (mg)	408	113	250	129	144	–	307	–	284
Copper (mg)	0.47	1.60	1.40	1.00	1.60	0.60	1.06	0.07	0.68
Manganese (mg)	5.49	0.60	0.60	0.68	1.10	0.96	1.15	0.51	2.29
Molybdenum (mg)	0.102	–	0.70	0.016	–	–	0.069	0.045	0.051
Zinc (mg)	2.3	1.4	2.4	3.7	0.7	3.0	3.1	1.3	2.7

Chromium (mg)	0.028	0.020	0.030	0.180	0.020	0.090	0.023	0.003	0.012
Sulfur (mg)	160	157	171	149	136	–	147	–	128
Chlorine (mg)	44	19	37	13	11	–	39	–	47
Essential amino acids (mg/g N)									
Arginine	300	290	220	250	270	–	300	480	290
Histidine	130	110	130	120	120	–	140	130	130
Lysine	220	190	140	110	150	–	190	230	170
Tryptophan	100	050	060	060	050	–	110	080	070
Phenyl alanine	310	310	420	330	430	–	290	280	280
Tyrosine	220	–	–	–	–	–	200	290	180
Methionine	210	160	180	180	180	–	150	150	090
Cystine	140	–	100	090	110	–	110	090	140
Threonine	240	150	190	190	200	–	240	230	180
Leucine	690	760	1040	760	650	–	750	500	410
Isoleucine	400	410	480	370	360	–	260	300	220
Valine	480	410	430	350	410	–	330	380	280

Source: Gopalan et al. (2009), Geervani and Eggum (1989).

percentage) of finger millet protein is 52 compared to 37 of sorghum and 63 of pearl millet (FAO, 1995). Finger millet contains higher levels of sulfur containing amino acids, namely, methionine and cystine, compared to milled rice (Table 1.1). The protein digestibility of finger millet is affected by the tannin content of the grain (Ramachandra, Virupaksha, & Shadaksharaswamy, 1977). However, Subrahmanyan, Narayana Rao, Rama Rao, and Swaminathan (1955) established that consumption of finger millet and pulse-based diet was sufficient to maintain a positive nitrogen (10.4% N), calcium (3.0% Ca), and phosphorus (8.7% P) balance in human adults. Doraiswamy, Singh, and Daniel (1969) also reported that supplementing finger millet diets with lysine or leaf (lucerne) protein improved the nutritional status, apparent protein digestibility, and N retention in children. Pore and Magar (1976) in their study reported a protein efficiency ratio (a measure of protein quality in terms of weight gain per amount of protein consumed) of 0.95 for B-11 variety of finger millet-based diet compared to 1.9 of control casein diet.

5.3. Fat

Mahadevappa and Raina (1978) reported 1.85–2.10% of total lipids in seven breeding varieties of finger millet. Finger millet lipids consist of 70–72% neutral lipids mainly triglycerides and traces of sterols, 10–12% of glycolipids, and 5–6% of phospholipids. On the whole, lipids contain 46–62% oleic acid, 8–27% linoleic acid, 20–35% palmitic acid, and traces of linolenic acid. Finger millet's fat content is lower compared to pearl millet, barnyard millet, little millet, and foxtail millet (Table 1.1), and this lower lipid content could be one of the factors contributing to the better storage properties of finger millet compared to other millets.

5.4. Micronutrients

Finger millet is exceptionally rich in calcium (344 mg%) compared to all other cereals and millets (eightfold higher than pearl millet) and contains 283 mg% phosphorus, 3.9 mg% iron (Gopalan et al., 2009), and many other trace elements and vitamins. Potassium content of finger millet is also high (408 mg%) compared to other cereals and millets (Table 1.1). High calcium finger millet varieties have also been reported elsewhere, and the "Hamsa" variety of finger millet was reported to contain much higher levels of calcium (660 mg%) (Umapathy & Kulsum, 1976). The phytic acid content of finger millet was lower than the levels present in common (proso) millet and foxtail millet and the values were in the range of 0.45–0.49 g% for different varieties of finger millet. The oxalate contents of finger millet were in the range of 29–30 mg% (Ravindran, 1991). Kurien, Joseph, Swaminathan, and Subrahmanyan (1959) reported that nearly 49% of total calcium content of finger millet is present in the husk. Sripriya, Antony, and Chandra (1997) reported that germination and fermentation of finger millet decreased the phytate content by 60% and improved bioavailability of minerals. Platel, Eipeson, and Srinivasan (2010) also reported increased bioaccessibility of minerals (iron, manganese) on malting of finger millet. Decortication of finger millet decreased the total mineral contents but increased the bioaccessibility of calcium, iron, and zinc, whereas popping of finger millet decreased the bioaccessibility of calcium but increased the bioaccessability of iron and zinc. Malting of finger millet increased the bioaccessibility of calcium, iron, and zinc (Rateesh, Usha, & Malleshi, 2012). A study conducted on 9- to 10-year-old girls showed that replacement of rice with finger millet diet apart from maintaining the positive nitrogen balance also improved calcium retention (Joseph, Kurien, Swaminathan, & Subrahmanyan, 1959).

Being rich source of calcium and iron, and the fact that the bioavailability can be improved by simple processing such as germination and fermentation, it should be considered as a good supplement for children and adolescents for improving bone health and hemoglobin.

6. PHYTONUTRIENTS/PHYTOCHEMICALS
6.1. Phenolics compounds

Varietal variations exist in phenolic content of finger millet (Chethan & Malleshi, 2007a; Ramachandra et al., 1977). Higher levels of phenolic compounds are reported in brown variety compared to white variety (Chethan & Malleshi, 2007a; Sripriya, Chandrasekharan, Murthy, & Chandra, 1996). Finger millet, in general, and the seed coat, in particular, contain several phytochemicals which may have health benefits (Chethan, 2008; Shobana, 2009; Shobana, Sreerama, & Malleshi, 2009). Finger millet is a very good source of variety of phenolic compounds (Chethan & Malleshi, 2007a, 2007b; Dykes & Rooney, 2007; Shobana, 2009; Shobana et al., 2009). Both free and bound forms of phenolic acids are reported in finger millet (Subba Rao & Muralikrishna, 2002). Caffeic acid is reported to decrease the fasting glycemia and attenuate the increase in plasma glucose in an intravenous glucose tolerance test in rats. It is also known to increase the glucose uptake in rat adipocytes and mice myoblasts. Catechin was found to improve the glucose tolerance in rats and quercitin inhibited glucose transport in transfected oocyte model and glucose absorption in rats (Matsumoto, Ishigaki, Ishigaki, Iwashina, & Hara, 1993; Scalbert, Manach, Morand, Remesy, & Jimenez, 2005). Finger millet contains these health potential polyphenols (Chethan & Malleshi, 2007b; Shobana et al., 2009). The impact of grain phenolics depends mainly on their bioavailability. Studies on the bioavailability of finger millet phenolics are scanty, and it is essential to study the nature of phenolic compounds in finger millet, their bioavailability, *in vivo* antioxidant functions, and the long-term effects of finger millet phenolics through human trials.

7. PROCESSING AND UTILIZATION

In India, usually finger millet is pulverized and the whole meal is utilized for the preparation of traditional foods, such as *roti* (unleavened flat breads), *kazhi* (finger millet balls), and *kanji* (thin porridge) (Fig. 1.1; Shobana, 2009). In addition to these traditional foods, finger millet is also processed to prepare popped, malted, and fermented products. The nonconventional products from finger millet are *papads* (rolled and dried preserved product), noodles, soup, etc. Recently, decorticated finger millet has been developed. The details of the different processing are presented in Fig. 1.2 (Malleshi, 2007; Rejaul, 2008; Shobana, 2009; Shobana & Malleshi, 2007; Ushakumari, 2009), and the nutrient composition of different finger millet products is presented in Table 1.2. A brief account of the nature of processing and the quality characteristics of the products are given below.

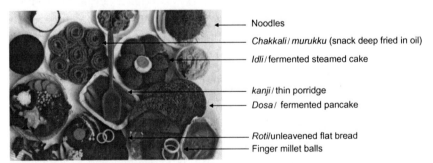

Noodles

Chakkali / murukku (snack deep fried in oil)

Idli / fermented steamed cake

kanji / thin porridge

Dosa / fermented pancake

*Roti/*unleavened flat bread
Finger millet balls

Figure 1.1 Prominent traditional Indian foods from finger millet. *Source: Photo courtesy Dr. N. G. Malleshi, Former Scientist, Department of Grain Science and Technology, Central Food Technological Research Institute, Mysore, India.* (For color version of this figure, the reader is referred to the online version of this chapter.)

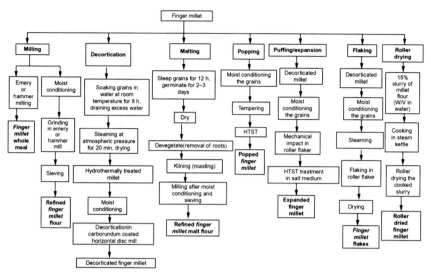

Figure 1.2 Different processing techniques adopted to develop products from finger millet. *Source: Malleshi (2007), Shobana (2009), Shobana and Malleshi (2007), Ushakumari (2009), Rejaul (2008).*

7.1. Milling

Finger millet kernel has a fragile endosperm with an intact seed coat, and due to these characteristic features, the grain cannot be polished and cooked in the grain or grit form similar to rice or other cereals. Hence the grain needs to be invariably pulverized or milled for preparation of flour. Generally, foods based on whole meal finger millet are darker, less attractive (Fig. 1.1). In view of this and to overcome the drawbacks, efforts had been made to prepare refined or seed coat free flour similar to white flour or "*maida*" (refined wheat flour) (Kurien & Desikachar, 1966). The refined flour is comparatively whiter and fairly free from the seed coat matter (SCM). However, the refined flour may have higher glycemic responses (GRs) compared to the wholemeal-based products owing to its predominant starch content and lower levels of dietary fiber (Shobana, 2009). Nutritionally also whole flour is superior to the refined flour in terms of its dietary fiber, vitamins, and mineral contents (Table 1.2).

7.2. Decortication

This is a very recent process developed for finger millet (Malleshi, 2006). The debranning or decortication methods followed for most of the cereals were not effective in the case of finger millet owing to the intactness of the seed coat with highly fragile endosperm. Hence to decorticate, finger millet

Table 1.2 Nutrient content of few of the finger millet products

Nutrients	Native finger millet	Refined finger millet flour	Malted finger millet refined flour	Hydrothermally processed finger millet (*dwb*)	Decorticated finger millet (*dwb*)	Expanded finger millet	Popped finger millet	Flaked finger millet	Finger millet toasted flakes[a]	Roller dried finger millet
Moisture (g)	9.8	11.0	8.2	–	–	10.0	3.4	8.5	3.1	2.9
Protein (g)	8.7	3.6	4.5	8.3	6.5	4.7	6.4	6.5	6.2	5.8
Fat (g)	1.5	0.9	0.6	1.3	0.8	0.7	0.9	0.7	0.6	0.6
Starch (g)	72.0	87.0	77.9	68.5	81.3	NA	69.1	74.1	77.0	78.9
Dietary fiber (g)										
Total	19.6	6.8	NA	18.1	10.3	11.3	NA	10.7	9.9	NA
Soluble	3.5	1.6		0.9	2.8	1.8		3.2	2.7	
Insoluble	16.1	5.2		17.2	7.5	9.5		7.5	7.2	
Ash (g)	2.2	1.1	0.9	1.8	1.1	1.1	2.0	1.0	1.5	1.5
Calcium (mg)	321	163	350	311	185	190.0	250	272	273	270
Phosphorus (mg)	201	106	190	160	111	NA	125	103	100	146

[a]Finger millet flakes toasted in salt as the heat transfer media.
NA, information not available.
Source: Shobana (2009), Shobana and Malleshi (2007), Ushakumari (2009); Rejaul (2008).

is hydrothermally processed (hydration, steaming, and drying) to harden the soft endosperm to enable it to withstand the mechanical impact during decortication (Fig. 1.2). The decorticated finger millet could be cooked as discrete grains similar to rice (Shobana & Malleshi, 2007). The SCM formed as the major by-product of the decortication process is a rich source of health beneficial phenolic compounds, minerals, and dietary fiber (Shobana, 2009). However, the glycemic property of decorticated finger millet (cooked in the discrete form similar to rice grains) is yet to be established. The SCM can be used as a source of fiber and micronutrients in food formulations.

7.3. Malting

Among the various tropical cereals, finger millet has good malting characteristics. Generally barley is preferred among cereals for malting both in brewing and in food industries. However, attempts were made as early as 1939 to study the malting characteristics of finger millet, and Sastri (1939) reported the conditions of producing good quality malt. The author also reported that the smallness of finger millet grain was advantageous for obtaining even germination as well as for kilning—the processes involved in the malt preparation. The details of the malting process are given in Fig. 1.2. The malt flour is a good source of amylases and is hence termed as "Amylase-rich food." Malt flour is a substitute to maltodextrin and can be blended with milk and spray dried to prepare infant food (Malleshi, 2007). During germination, the amylases partially hydrolyze the starch to lower molecular weight carbohydrates such as oligo- and disaccharides, and thus the malt flour has reduced water holding capacity and thus high energy density. Due to this, the refined finger millet malt flour has scope for utilization in infant foods (Malleshi & Gokavi, 1999), weaning foods (Malleshi & Desikachar, 1982), and enteral foods (Malleshi, 2002; Malleshi & Chakravarthy, 1994). The millet malt flour has also been utilized in milk-based beverages, confectionary, and cakes (Desai, Kulkarni, Sahoo, Ranveer, & Dangde, 2010). Malted cereals contain highly digestible carbohydrates and may exhibit higher GRs (Lakshmi Kumari & Sumathi, 2002; Sumathi, Vishwanatha, Malleshi, & Venkat Rao, 1997) and hence may not be suitable for other metabolic conditions.

7.4. Popping

This is one of the important processing techniques widely used to prepare ready-to-eat products, which involves high-temperature short time treatment (HTST) to finger millet using sand as heat transfer media, where finger millet starch gelatinizes and the endosperm bursts open. Popped finger millet possesses a highly desirable flavor and aroma. It is used as a snack after seasoning with spices and condiments. Popped finger millet flour is commonly known as *hurihittu* in the state of Karnataka, India. It is a whole-grain product rich in macronutrients, micronutrients, dietary fiber, and usually mixed with vegetable or milk protein sources such as popped bengal gram, milk powder, and oil seeds, sweetened with jaggery or sugar to prepare ready-to-eat nutritious supplementary food (Malleshi, 2007). In recent days, finger millet is puffed using an air popper device also which is superior (in terms of quality with no sand and low salt in the product) compared to the product prepared using sand or salt as heat transfer media. Recently, expanded finger millet has also been developed from the decorticated finger millet (Ushakumari, Rastogi, & Malleshi, 2007). The dietary fiber content of the expanded finger millet is lower than that of popped finger millet as it is prepared from decorticated finger millet which is devoid of seed coat.

Fermented beverages from finger millet are also common. A traditional mild–alcoholic beverage prepared from finger millet (Kodo ko jaanr) is consumed in Eastern Himalayan areas of the Darjeeling hills and Sikkim in India, Nepal, and Bhutan (Thapa & Tamang, 2004). The relatively newer food products which are currently being explored are noodles; vermicilli prepared either out of finger millet alone or in combination with refined wheat flour (Shukla & Srivastava, 2011; Sudha, Vetrimani, & Rahim, 1998); pasta products (Krishnan & Prabhasankar, 2010); *halwa* mixes (a sweet dish prepared with flour, sugar, and clarified butter) and composite mixes (Itagi & Singh, 2011; Itagi, Singh, Indiramma, & Prakash, 2011); *papads* (flattened and dried dough products which are toasted or deep fried and used as adjuncts with a meal) (Kamat, 2008; Vidyavati, Mushtari, Vijayakumari, Gokavi, & Begum, 2004); roller dried finger millet-based soup mixes (Guha & Malleshi, 2006); bakery products such as muffins (Jyotsna, Soumya, Indrani, & Venkateswara Rao, 2011); bread and biscuits (Krishnan, Dharmaraj, Sai Manohar, & Malleshi, 2011; Saha et al., 2011; Singh, Abhinav, & Mishra, 2012); and complementary foods (Stephen et al., 2002) are also being prepared and marketed in selected markets. Breads from millet-based composite flours, wheat in combination with finger millet, barnyard millet, and proso millet, were also prepared (Singh et al., 2012). Recently, attempts have been made for fortification of finger millet flour with iron and zinc. Fortication of iron (in the form of ferrous fumarate and ferric pyrophosphate) in finger millet flour with EDTA and folic acid as cofortificants significantly increased the bioaccessibility of iron from the fortified flours (Tripathi & Platel, 2011). And inclusion of EDTA as a cofortificant along with the zinc (zinc oxide or zinc stearate) significantly enhanced the bioaccessibility of zinc from the fortified flours (Tripathi & Platel, 2010). Studies on the double fortification of finger millet flour with zinc and iron with disodium EDTA as a cofortificant are also available in literature (Tripathi, Platel, & Srinivasan, 2012). Raghu and Bhattacharya (2010) computed the optimum conditions suitable for developing a flattened product and bread spread from finger millet. The authors reported that treating finger millet flour dough with bacterial α-amylase as an option to modify the attributes of finger millet flour dough related to processing and product development. They also added that enzyme-treated doughs were more sticky and had lower firmness compared to the untreated samples.

In summary, finger millet whole flour has considerable versatility and could be used for the preparation of traditional and contemporary food products in lieu of rice and wheat. If scientific evidence for health benefits is made available, finger millet products would have better market potentials.

8. HEALTH BENEFITS OF FINGER MILLET

Several *in vitro* (Table 1.3a) and *in vivo* studies (animal) (Table 1.3b) have been conducted to explore the health benefits of finger millet.

Several studies are available on the antioxidant properties (Chandrasekara & Shahidi, 2011; Hegde & Chandra, 2005; Sripriya et al., 1996; Subba Rao & Muralikrishna, 2002; Varsha et al., 2009; Veenashri & Muralikrishna, 2011) and antimicrobial properties of finger millet (Antony et al., 1998; Chethan & Malleshi, 2007a; Varsha et al., 2009). Production of statins (antihypocholestrolemic metabolites) from finger millet was attempted by Venkateswaran and Vijayalakshmi (2010). α-Glucosidase inhibitors play a vital role in the clinical management of postprandial hyperglycemia, and Shobana et al. (2009) established the α-glucosidase and pancreatic α-amylase inhibitory properties of finger millet phenolic extract, whereas Chethan et al. (2008) and Chethan (2008) in their study indicated that finger millet phenolics are inhibitors of aldose reductase and snake venom phospholipases (PLA_2). Protein glycation is one of the complications of diabetes, and protein glycation inhibitors are helpful in the management of this complication. Methanolic extracts of finger millet were found to exhibit protein glycation inhibitory properties (Hegde et al., 2002; Shobana, 2009).

Animal studies on the health beneficial aspects of finger millet feeding are also available in the literature. As early as 1975, Tovey et al. (1975) reported

Table 1.3 Some of the known health beneficial properties of finger millet

References	Functional property	Method	Key findings
(a) in vitro studies			
Chandrasekara and Shahidi (2011)	Antioxidant property	Singlet oxygen inhibition, oxygen radical absorbance capacity (ORAC), DPPH radical scavenging assay by electron paramagnetic resonance (EPR) spectroscopy, β–carotene linoleate model system	• Insoluble bound phenolic fractions of finger millet showed lower ORAC than that of free phenolic fraction due to their lower total phenolic content • The higher antioxidant capacity of the free phenolic fraction of finger millet—attributed to the high total phenolic content as well as flavonoids such as catechin, gallocatechin, epicatechin, and procyanidin dimmer contents detected in the free fraction of finger millet
Veenashri and Muralikrishna (2011)	Antioxidant property	1,1-Di phenyl-2 picryl-hydrazyl (DPPH), β–carotene linoleate emulsion, and ferric reducing antioxidant power (FRAP) assays	• Xylo-oligosaccharides (XO) from finger millet bran exhibited higher antioxidant activity of about 70% at 60 μg concentration, which is more than the antioxidant activity exhibited by rice, maize, and wheat bran XO mixtures (70% at 1000 μg concentration) in DPPH and FRAP assays
Venkateswaran and Vijayalakshmi (2010)	Production of antihypercholestrolemic metabolites	Finger millet (native and germinated)—used as substrates for solid state fermentation of *Monascus purpureus* at 28 °C for 7 days using 2% seed medium as inoculum for the production of its metabolites	• Germinated finger millet yielded higher total statin production of 5.2 g/kg dry weight with pravastatin and lovastatin content of 4.9 and 0.37 g/kg dry weight, respectively

Reference	Activity/Property	Method	Findings
Shobana et al. (2009)	Inhibition of intestinal α-glucosidase and pancreatic amylase	Enzyme kinetic studies	• Inhibition of intestinal α-glucosidase and pancreatic amylase by phenolics from finger millet seed coat matter—IC$_{50}$ values 16.9 and 23.5 µg of phenolics • Noncompetitive type of inhibition for both the enzymes
Shobana (2009)	Antiprotein (albumin) glycation property	Fructose–bovine serum albumin model system	• Finger millet seed coat polyphenols (MSCP)—effective inhibitors of fructose-induced albumin glycation • Better restoration of tryptophan fluorescence (an index of inhibition of fructose-induced glycation of albumin) by MSCP (45 µg) compared to 15 mM aminoguanidine, a standard glycation inhibitor • 41% decrease in the AGE (advanced glycation end product) formation by 45 µg of MSCP compared to 54% decrease produced by 15 mM aminoguanidine
Varsha, Urooj, and Malleshi (2009)	Antibacterial and antifungal activity	Disc diffusion method, spore germination method	• Higher antimicrobial activity for polyphenol extract from finger millet seed coat compared to the polyphenol extract from finger millet whole flour against *Bacillus cereus* and *Aspergillus flavus*

Continued

Table 1.3 Some of the known health beneficial properties of finger millet—cont'd

References	Functional property	Method	Key findings
Varsha et al. (2009)	Antioxidant activity	Reducing power assay, β-carotene–linoleic acid assay	• High reducing power for seed coat polyphenol extract compared to the polyphenol extract from finger millet whole flour • Higher antioxidant activity for finger millet seed coat polyphenol extract (86%) compared to polyphenol extract from finger millet whole flour (27%)
Chethan, Dharmesh, and Malleshi (2008)	Aldose reductase enzyme inhibitory property	Enzyme kinetic studies	• Finger millet seed coat polyphenols inhibits aldose reductase extracted from cataracted human eye lenses by noncompetitive mode of inhibition • Quercitin from finger millet was more effective in inhibition with IC_{50} of 14.8 nM
Chethan (2008)	Inhibition of phospholipases A_2 (PLA_2) from snake venom (*Naja naja*)	Enzyme inhibitory studies	• Crude polyphenol extract from finger millet was a potent inhibitor of PLA_2 (IC_{50} 83.2 μg/ml) • Gallic acid and quercitin from finger millet was more potent inhibitors of PLA_2 (IC_{50} 62.3 and 16.8 μg/ml, respectively)
Chethan (2008) and Chethan and Malleshi (2007a)	Inhibition of pathogenic bacterial strains	Agar diffusion assay	• Antimicrobial activity of finger millet polyphenols on pathogenic bacterial strains: *Escherichia coli*, *Staphylococcus aureus*, *Listeria monocytogenes*, *Streptococcus*

Reference	Property	Method	Findings
			• pyogenes, *Pseudomonas aeruginosa, Serrtia marcescens, Klebsiella pneumonia*, and *Yersinia enterocolitica* • Quercitin fractionated from finger millet inhibited the growth of all the pathogenic bacteria • Gallic, caffeic, protocatechuic, *para*-hydroxy benzoic acid showed their antimicrobial potentials restricted to only few bacterial strains
Hegde, Chandrakasan, and Chandra (2002)	Inhibition of collagen glycation and cross-linking	Phenol sulfuric acid method, pepsin digestion, and cyanogen bromide peptide mapping	• Methanolic extracts of finger millet inhibited both glycation and cross-linking of collagen
Subba Rao and Muralikrishna (2002)	Antioxidant activity	β-Carotene–linoleic acid assay	• Free phenolic acid content increased upon malting • Bound phenolic acid content decreased upon malting • Higher antioxidant activity for free phenolic acid mixture compared to that of bound phenolic acid mixture from finger millet • Malting influences antioxidant property of finger millet
Antony, Moses, and Chandra (1998)	Antimicrobial properties		• Inhibition of *Salmonella typhimurium* and *Escherichia coli* by fermented flour of finger millet

Continued

Table 1.3 Some of the known health beneficial properties of finger millet—cont'd

References	Functional property	Method	Key findings
Sripriya et al. (1996)	Free radical scavenging	Electron spin resonance (ESR) spectrometry	• The phenolic acid content of brown finger millet 96% higher compared to white variety • Germination and/or fermentation processes decreased the free radical quenching action of brown finger millet • Brown finger millet found to be more efficient in free radical quenching
(b) in vivo studies			
Shobana, Harsha, Platel, Srinivasan, and Malleshi (2010)	Blood glucose lowering effect	Streptozotocin-induced diabetic rat model fed with diet containing 20% finger millet seed coat matter for 6 weeks	• 39% reduction in the fasting blood glucose levels in diabetic experimental group of animals compared to the diabetic controls • Lower glycosylated hemoglobin levels (4.21%) compared to diabetic controls (6.31%)
	Decreases AGE (advanced glycation end product) formation		• 20% lower AGE products at the end of the study compared to the diabetic controls
	Delay of onset of cataractogenesis		• Lower aldose reductase enzyme activity in the eye lens compared to the diabetic control group • Very mild lenticular opacity and posterior subcapsular cataract compared to the significant mature cataract in the diabetic control group of animals

	Nephroprotective properties	• Decrease in the urinary volume and excretion of urinary metabolites (glucose, protein, urea, and creatinine) compared to the diabetic controls • Normal glomerular, tubular structures and absence of mucopolysaccharide depositions in the kidney compared to the shrunken glomeruli, tubular clarifications, vacuolizations, and mucopolysaccharide depositions in the kidney of diabetic controls
	Cholesterol lowering	• Lower serum cholesterol and triacyl glycerol levels (43% and 62%, respectively) compared to diabetic controls • Lower atherogenic index (1.82) compared to the diabetic controls (5.3)
Tatala, Ndossi, Ash, and Mamiro (2007)	Improvement on hemoglobin status in children	The effect of feeding germinated finger millet–based food supplement on hemoglobin status—assessed in children in rural Tanzania. The food consisted of finger millet flour, kidney beans, ground peanuts, and dried mangoes (75:10:10:5), diets were supplemented to children for 6 months	• Supplementing infants with the germinated finger millet–based food showed a general improvement on hemoglobin status

Continued

Table 1.3 Some of the known health beneficial properties of finger millet—cont'd

References	Functional property	Method	Key findings
Lei, Friis, and Michealsen (2006)	Natural probiotic treatment for diarrhea	Finger millet drink fermented by lactic acid bacteria used as a therapeutic agent among Ghanaian children with diarrhea (below 5 years). Children randomized to two groups, both groups received treatment for diarrhea and the intervention group in addition received up to 300-ml fermented millet drink (KSW) daily for 5 days	• No effects of the intervention were found with respect to stool frequency, stool consistency, and duration of diarrhea • Greater reported well-being 14 days after the start of the intervention with KSW • No effect of KSW on diarrhea could be because many were treated with antibiotics which could have affected the lactic acid bacteria, or the lactic acid bacteria in KSW had no probiotic effects • Effect after two weeks could be due to a preventing effect of KSW on antibiotic-associated diarrhea which could help reducing persistent diarrhea
Hegde, Rajasekaran, and Chandra (2005)	Cholesterol lowering	Alloxan-induced diabetic rat model, finger millet incorporated diets (55% in the basal diet) fed for 28 days	• 13% reduction in serum cholesterol levels compared to the controls
	Rat tail collagen glycation inhibitory property		• Rat tail tendon collagen glycation was only 40% in the finger millet fed rats compared to control group of animals
	Blood glucose lowering property		• 36% reduction in blood glucose levels
	Antioxidant property		• Levels of enzymatic (catalase, superoxide dismutase, glutathione peroxidase, and glutathione reductase) and nonenzymatic

References	Property	Study model	Observations
			• antioxidants (glutathione, vitamin E and C) restored to normal levels • Reduced levels of lipid peroxides
Hegde, Anitha, and Chandra (2005)	Wound healing property	Excision wound—rat model, finger millet flour (300 mg)—aqueous paste applied topically once daily for 16 days	• Significant increase in protein, collagen, and decrease in lipid peroxides • 90% rate of wound contraction compared to 75% for control untreated rats • Complete closure of wounds after 13 days compared to 16 days for untreated rats
Rajasekaran, Nithya, Rose, and Chandra (2004)	Wound healing property	Excision skin wound—hyperglycemic (alloxan-induced) rat model, 4 weeks' feeding	• Increased expression of NGF (nerve growth factor) • Epithelialization, increased synthesis of collagen, activation of fibroblasts, and mast cells • Control of blood glucose levels, improved antioxidant status, hastened dermal wound healing
Pore and Magar (1976)	Cholesterol lowering	Male albino rats fed with ragi diet for 8 weeks	• Lower serum cholesterol (65 mg/dl) compared to casein diet fed animals (95 mg/dl)
Tovey, Jayaraj, and Clark (1975)	Antiulcerative property	Experimental ulceration following pyloric ligation in female albino rats fed with ragi incorporated stock diet for 2 weeks	• Offered protection against mucosal ulceration

the antiulcerative potential of finger millet through animal feeding trials and Pore and Magar (1976) reported the cholesterol lowering properties of finger millet by animal feeding. The other health beneficial aspects of finger millet feeding, namely, the glucose lowering, cholesterol lowering, nephroprotective properties, antioxidant properties, wound healing properties, and anticataractogenesis properties of finger millet were reported by several authors (Hegde, Anitha, et al., 2005; Hegde, Rajasekaran, et al., 2005; Rajasekaran et al., 2004; Shobana et al., 2010). Improvement on the status of hemoglobin in children on feeding finger millet-based food was reported by Tatala et al. (2007). Whereas fermented finger millet drink as a natural probiotic treatment for diarrhea was reported by Lei et al. (2006). Even though there are many *in vitro* and *in vivo* studies on the health benefits of finger millet, there are very few human studies to the best of our knowledge.

9. *IN VITRO* STUDIES ON THE CARBOHYDRATE DIGESTIBILITY OF FINGER MILLET

Studies on the type of starch, its digestibility, and crystallinity demonstrated that the degree of crystallinity of finger millet and the amount of heat flow required to gelatinize starch was much higher as compared to rice. The molecular weight of the human salivary amylase digests of the finger millet starch was higher than that of rice starch digests (Mohan, Anitha, Malleshi, & Tharanathan, 2005). Further studies indicate that finger millet starch is the most difficult to be hydrolyzed in *in vitro* by fungal α-amylase (Singh & Ali, 2006). These results indicate a higher degree of crystallinity and slightly lower digestibility of finger millet starch by the digestive enzymes in *in vitro*.

Roopa and Premavalli (2008) studied the effect of processing on the starch fractions in finger millet and reported that puffing resulted in a mild increase in the rapidly digestible starch and decrease in the slowly digestible starch fractions compared to native finger millet flour, and the authors also reported that resistant starch (RS) content decreased during puffing. The study also indicated that the pressure cooking and roasting processes increased the RS fraction in finger millet compared to other processes. The finger millet product prepared by roasting contained the highest amount of RS compared to other products (3.1%).

Several *in vitro* and *in vivo* studies on the carbohydrate digestibility and glycemic properties of finger millet foods indicated that the rate of starch hydrolysis and glucose release (digestibility index, DI) are affected by degree of gelatinization (DG), added ingredient components and accompaniments (Roopa, Urooj, & Puttaraj, 1998; Sharavathy, Urooj, & Puttaraj, 2001). Finger millet *puttu* (steamed product made out of finger millet flour and consumed with grated coconut and sugar) registered a lower DG as compared to rice *puttu* and other finger millet and rice preparations (*roti, dosa,* dumpling) (Roopa et al., 1998). It is to be noted that wheat *chapathis* have better DI as compared to finger millet *roti* (Urooj & Puttaraj, 1999).

Urooj, Rupashri, and Puttaraj (2006) reported a high DG and DI for dumplings compared to the *roti* preparations from finger millet and sorghum. The study also reported a higher GI for dumplings compared to *roti* preparations. This clearly shows that DG affects the GI of the food. Similar studies on rice have indicated that rice with lower DG elicited a lower glucose and insulinemic responses compared to the rice with higher DG (Jung et al., 2009). Reports on potato starch indicate that the DG of starch strongly affects its digestibility *in vitro* (enzymatic digestion) and may influence the postprandial GR (Parada & Aguilera, 2008). The digestibility or hydrolysis of starch is considerably influenced by accompaniments. *Puttu* (a steamed cereal flour) with coconut and sugar resulted in higher DI as compared to the cereal foods with *chutney* or *sambar* as accompaniments (Roopa et al., 1998). In short, the food matrix influences the carbohydrate digestibility.

It needs to be observed that repeated cooking and cooling for several times for both rice and finger millet flour suspensions led to increased levels of RS (Mangala, Malleshi, Mahadevamma, & Tharanathan, 1999; Mangala, Ramesh, Udayasankar, & Tharanathan, 1999). The RS isolated from five-cycle autoclaved finger millet flour was found to be a linear α-1,4-D-glucan, probably derived from a retrograded amylose fraction of finger millet starch. The undigested material recovered from the ileum of rat intestine fed with processed finger millet flour exhibited a close similarity (almost comparable molecular weight) in some of its properties to that of RS isolated by an *in vitro* method (Mangala, Ramesh, et al., 1999). Lower GRs for cooled potatoes compared to hot potatoes are reported elsewhere (Nadine, Nada, & Nahla, 2004). Aarthi, Urooj, and Puttaraj (2003) reported an RS content of 4.5% in finger millet *roti*, and Platel and Shurpalekar (1994) reported an RS content of 1.79 g% (on dry weight basis) in pressure cooked finger millet. Foods containing RS have reduced digestible energy (hence reduced calorie density) compared to the foods containing readily digestible starch. Apart from these, RS produces lower glycemic and insulinemic responses and is known to be beneficial for individuals with diabetes and obesity (Birkett, 2007).

10. GLYCEMIC RESPONSE (GR) STUDIES ON FINGER MILLET (HUMAN STUDIES)

As early as 1957, Ramanathan and Gopalan (1957) reported low GR to cooked finger millet and finger millet starch isolates compared to milled rice, parboiled rice, and wheat diet when very little was known about the factors that influence the glycemic response. Though there were other studies on the GR to finger millet preparations, compared to other cereals and millets (Chitra & Bhaskaran, 1989; Kavita & Prema, 1995; Kurup & Krishnamurthy, 1992; Lakshmi Kumari & Sumathi, 2002; Mani, Prabhu, Damle, & Mani, 1993; Shobana, Ushakumari, Malleshi, & Ali, 2007; Shukla & Srivastava, 2011; Urooj et al., 2006), it is difficult to draw any conclusion as the number of subjects participated in the study was either too small or, even with sufficient sample size, the methodology adopted was inconsistent compared to the internationally valid GI protocols. Most of the studies were conducted with small number of individuals, and measurement of the glycemic response was done using venous samples instead of capillary blood samples as suggested by the new valid protocol. Capillary samples are generally preferred over venous samples as the rise in blood glucose values in response to foods is higher than in venous plasma and less variable in capillary blood samples as compared to that of venous plasma glucose. Hence, the differences between foods are larger and easier to detect statistically using capillary blood glucose (FAO/WHO, 1998; Venn & Green, 2007). In addition, the available carbohydrate content of the test foods was not estimated directly then, as the standardized protocol became available recently and the method of the food preparation and accompaniments were not adequately reported (potential factors to influence GI) in these studies.

Geetha & Easwaran (1990) in their study fed finger millet incorporated (20%) breakfast preparations, namely, *idli* (fermented, steamed rice cake), *dosai* (fermented rice pan cake), *chapathi* (unleavened flat bread), *sevai* (rice string hoppers), and *kozhukattai* (steamed rice balls) to NIDDM subjects (n = 6) for a month. The study showed a significant decrease in the postprandial blood glucose after finger millet administration for a month, though the limitation was small sample size.

Lakshmi Kumari and Sumathi (2002) also reported that malting and fermentation processes increase the carbohydrate digestibility of finger millet and attributed the higher GR of germinated finger millet dosa (fermented) and germinated finger millet *roti* compared to the normal whole finger millet *dosa* and whole finger millet *roti* possibly due to the conversion of starch to lower molecular weight dextrins and maltose during germination. The authors reported a higher GR for finger millet *dosa* compared to finger millet *roti* and attributed their results to the wet grinding and fermentation of batter prior to *dosa* preparation leading to starch gelatinization and enhanced rate of starch digestion in finger millet *dosa* compared to finger millet *roti*.

Decorticated (polished) finger millet in the grain form as a formulation with legumes gave a higher GI (93.4) as compared to wheat-based formulations with legumes (55.4) (Shobana et al., 2007). Lower GI for finger millet incorporated noodles compared to refined wheat flour-based noodles was reported by Shukla and Srivastava (2011).

Patel, Dhirawani, and Dharne (1968) did not observe any benefit in the treatment of diabetes when the diets of eight male diabetic subjects (40–80 years of age) with duration of diabetes from 1 day to 13 years were changed from rice to finger millet as the staple grain in their diets for 8–10 weeks. The recent report on the nutritional management of diabetes with finger millet is still not very conclusive, and hence studies for prolonged periods are needed to draw meaningful conclusions (Kamala et al., 2010; Pradhan, Nag, & Patil, 2010).

This drawback is not only in the case of studies on finger millet, but it also applies to the studies on GI of other millets. Mani et al. (1993) indicated that the GI of finger millet in the form of roasted bread (unleavened flat bread) (104) was higher than the GI values obtained for pressure cooked *varagu* (kodo millet) (68) and roasted breads from *bajra* (pearl millet) (55) and *jowar* (sorghum) (77). While Pathak, Srivastava, and Grover (2000) determined the GI of *dhokla* (fermented and steamed cakes), *uppma* (cooked roasted semolina seasoned with vegetables and spices), and *laddu* (an Indian sweet ball) prepared from a blend of foxtail and barnyard millet with fenugreek and legumes and reported a GI of 34.96 for *dhokla*, 23.52 for *laddu*, and 17.60 for *uppma*. Vijayakumar, Mohankumar, and Srinivasan (2010) in a recent study reported a GI of 84.8 for barnyard and kodo millet incorporated noodles compared to 94.6 for the branded noodle prepared from *maida* (refined wheat flour) in 10 normal subjects. There was no indication on the method of estimation of available carbohydrates, and the capillary blood glucose response was determined at 1-h intervals (0, 1, 2, and 3 h). The GI reported in the literature for foxtail millet incorporated biscuit was 50.8 compared to 68.0 for both barnyard millet incorporated biscuits and control refined flour wheat biscuits (Anju & Sarita, 2010). Ugare, Chimmad, Naik, Bharati, and Itagi (2011) in their study reported a GI of 50.0 and 41.7 for the dehulled and heat-treated barnyard (dehulled) millet, respectively. Inspite of several studies on the GI of millets, due to the outdated methodology

Table 1.4 A sample meal plan with "finger millet"-based preparations

Meal	Menu	Description
Breakfast	Finger millet steamed cake	Fermented steamed cake prepared from finger millet and black gram dhal
	Tomato and onion sauce	Ground paste of tomatoes and onions seasoned with spices
	Lentil sauce	Spicy lentil curry in tamarind sauce
Lunch	Finger millet balls/ unleavened flat bread	Gelatinized stiff porridge made into balls with whole finger millet flour/finger millet flour based unleavened flat bread
	Egg plant sauce	Spicy lentil curry in tamarind sauce with egg plant
Evening tea	Tea	With milk
	Finger millet crisps	Low fat extruded snack prepared from finger millet flour
Dinner	Finger millet fermented pan cake	Fermented pancake made with ground batter of finger millet and black gram dhal
	Vegetable stew	Vegetable stew with spices

adopted in these studies, it is very difficult to draw meaningful interpretations on the GI of other millets too.

However, it is a general belief that lower glucose responses are obtained with unrefined finger millet-based products and health professionals tend to advice finger millet preparations for treatment of both obesity and diabetes. There are many traditional (Fig. 1.1) and contemporary foods that could be prepared with whole finger millet meal. These foods could be used in place of staple foods such as rice and wheat. Table 1.4 lists such a sample meal plan using finger millet-based preparations. However the GI testing of the individual preparations should be carried out before recommending these preparations for population with diabetes. Judicious selection of accompaniments should also be made for these preparations to lower the overall meal glycemic response.

11. GAPS IN THE KNOWLEDGE AND FUTURE DIRECTIONS FOR RESEARCH

Though finger millet inherently possess components that are likely to lower GI, various methodologies and nonstandard protocols adopted for the determination of GI in most of the studies resulted in GI values ranging from low to high GI categories. Moreover, the available carbohydrate (glycemic carbohydrates, carbohydrates available for metabolism) content of the foods was not estimated by direct measurement and most of the values were calculated from food composition data that give carbohydrate values "*by difference*" method (Gopalan et al., 2009). The updated GI methodology published elsewhere (Brouns et al., 2005; FAO/WHO, 1998; Wolever, Jenkins, Jenkins, & Joss, 1991) recommends volunteers aged between 18 and 45 years and inclusion of at least 10 subjects for the determination of GI of a food. Capillary blood glucose is preferred over venous sample to give the greatest sensitivity to the food's GI, whereas most of the studies cited mostly used venous samples. In addition, the current GI methodology guidelines strongly recommend direct measurement (by enzymatic procedure) of available carbohydrates from foods. Imprecise estimation of the available carbohydrate content of foods can lead to over or underestimation of GI values of the food (Brouns et al., 2005).

The method of processing, DG, particle size, food ingredients, accompaniments, and the form in which the food is consumed are known to affect GI (Brand Miller, Wolever, Foster-Powell, & Colagiuri, 2007). Hence, the glycemic response to finger millet-based preparations should be determined in isolation and also along with usual accompaniments using validated protocol to understand its glycemic properties.

12. CONCLUSION

Whole-grain consumption is associated with reduced risk of CVD and type 2 diabetes; hence, consumption of whole meal-based finger millet products may be desirable due to the protective role of SCM that could contribute varied health benefits. Finger millet products processed suitably to lower their GI in synergy with accompaniments rich in vegetables and pulses may help in prevention or control of chronic diseases in general and diabetes in particular. The published literature shows a wide range of GI values for foods from finger millet (45–104), despite methodological limitations and due to processing attributes like particle size, germination, hydrothermal processing, decortication, roasting, cooking in the form of dumpling, etc. The presence of higher levels of dietary fiber and other protective healthy nutrients in the whole meal-based finger millet preparations may still help in the nutritional management of diabetes. Hence to enjoy the benefit of the functional constituents, it is imperative to identify judicious processing methods for finger millet preparations to obtain maximum health benefits from the millet through appropriate research. GI testing of finger millet preparations as well as short- and long-term intervention trials using judiciously processed finger millet may be helpful in understanding the health benefits of finger millet before appropriate recommendations could be made for prevention and management of chronic diseases such as diabetes. Studies in this direction are already in progress at Madras Diabetes Research Foundation, Chennai, India.

Declaration of interests: S. S. wrote the first draft of the manuscript. K. K., V. S., N. G. M., R. M. A., L. P., and V. M.. reviewed the manuscript and contributed to the revision and finalization of the manuscript. All authors declared that they have no duality of interest associated with this manuscript.

REFERENCES

Aarthi, A., Urooj, A., & Puttaraj, A. (2003). Starch digestibility and nutritionally important starch fractions in cereals and mixtures. *Starch/Starke*, *55*, 94–99.

Achaya, K. T. (1992). *Indian food: A historical companion*. USA: Oxford University Press.

Achaya, K. T. (2009). *The illustrated food of India A–Z*. New Delhi, India: Oxford University Press.

Anju, T., & Sarita, S. (2010). Suitability of Foxtail millet (*Setaria italica*) and Barnyard millet (*Echinochloa frumentacea*) for development of low glycemic index biscuits. *Malaysian Journal of Nutrition*, *16*(3), 361–368.

Antony, U., Moses, L. G., & Chandra, T. S. (1998). Inhibition of *Salmonella typhimurium* and *Escherichia coli* by fermented flour of finger millet. *World Journal of Microbiology and Biotechnology*, *14*(6), 883–886.

Augustin, L. S., Franceschi, S., Jenkins, D. J. A., Kendall, C. W. C., & Vecchia, C. L. (2002). Glycemic index in chronic disease: A review. *European Journal of Clinical Nutrition*, *56*, 1049–1071.

Backes, A. C., Abbasi, F., Lamendola, C., McLaughlin, T. L., Reaven, G., & Palaniappan, L. P. (2008). Clinical experience with a relatively low carbohydrate, calorie-restricted diet improves insulin sensitivity and associated metabolic abnormalities in overweight, insulin resistant South Asian Indian women. *Asia Pacific Journal of Clinical Nutrition*, *17*(4), 669–671.

Birkett, A. M. (2007). Resistant starch. In C. J. K. Henry (Ed.), *Novel food ingredients for weight control* (pp. 174–197). Washington, DC/England: Woodhead Publishing Limited/CRC Press.

Brand Miller, J., Wolever, T. M. S., Foster-Powell, K., & Colagiuri, S. (2007). *The new glucose revolution* (3rd ed.). New York: Marlowe & Company.

Brouns, F., Bjorck, I., Frayn, K. N., Gibbs, A. L., Lang, V., Slama, G., et al. (2005). Glycemic index methodology. *Nutrition Research Reviews*, *18*(1), 145–171.

Chandrasekara, A., & Shahidi, F. (2011). Determination of antioxidant activity in free and hydrolyzed fractions of millet grains and characterization of their phenolic profiles by HPLC-DAD-ESI-MSn. *Journal of Functional Foods*, *3*, 144–158.

Chethan, S. (2008). Finger millet (*Eleusine coracana*) seed polyphenols and their nutraceutical potential. Ph.D. Thesis. University of Mysore, Mysore, India.

Chethan, S., Dharmesh, M. S., & Malleshi, N. G. (2008). Inhibition of aldose reductase from cataracted eye lenses by finger millet (*Eleusine coracana*) polyphenols. *Bioorganic & Medicinal Chemistry*, *16*, 10085–10090.

Chethan, S., & Malleshi, N. G. (2007a). Finger millet polyphenols: Characterization and their nutraceutical potential. *American Journal of Food Technology*, *2*(7), 582–592.

Chethan, S., & Malleshi, N. G. (2007b). Finger millet polyphenols: Optimization of extraction and the effect of pH on their stability. *Food Chemistry, 105*, 862–870.

Chitra, K. V., & Bhaskaran, T. (1989). Glycemic response of diabetics to selected cereals administered in different forms. *The Indian Journal of Nutrition and Dietetics, 26*, 122–128.

de Munter, J. S. L., Hu, F. B., Spiegelman, D., Franz, M., & van Dam, R. M. (2007). Whole grain, bran, and germ intake and risk of type 2 diabetes: A prospective Cohort study and systematic review. *PLoS Medicine, 4*(8), 1385–1395.

Desai, A. D., Kulkarni, S. S., Sahoo, A. K., Ranveer, R. C., & Dangde, P. B. (2010). Effect of supplementation of malted *ragi* flour on the nutritional and sensorial characteristics of cake. *Advance Journal of Food Science and Technology, 2*(1), 67–71.

Dixit, A. A., Azar, K. M. J., Gardner, C. D., & Palaniappan, L. P. (2011). Incorporation of whole, ancient grains into a modern Asian Indian diet to reduce the burden of chronic disease. *Nutrition Reviews, 69*(8), 479–488.

Doraiswamy, T. R., Singh, N., & Daniel, V. A. (1969). Effects of supplementing ragi (*Eleusine coracana*) diets with lysine or leaf protein on the growth and nitrogen metabolism of children. *British Journal of Nutrition, 23*, 737–743.

Dykes, L., & Rooney, L. W. (2007). Phenolic compounds in cereal grains and their health benefits. *Cereal Foods World, 52*(3), 105–111.

Food and Agricultural Organisation (FAO) of the United Nations. (1995). Sorghum and millets in human nutrition (FAO Food and Nutrition Series, No. 27). ISBN 92-5-103381-1.

FAO/WHO, (1998). Carbohydrates in human nutrition: Report of joint FAO/WHO expert consultation. *FAO Food and Nutrition, 66*, 1–140.

Fuller, D. Q. (2002). Fifty years of archeaobotanical studies in India: Laying a solid foundation. In S. Settar & R. Korisettar (Eds.), *Indian archealogy in retrospect Archeology and interactive disciplines: Vol. III.* (pp. 247–364). Delhi: Manohar.

Fuller, D. Q. (2003). African crops in prehistoric South Asia: A critical review. In K. Neumann, A. Butler & S. Kahlheber (Eds.), *Food, fuel and fields. Progress in African archaeobotany* (pp. 239–271). Koln: Heinrich-Barth Institute.

Geervani, P., & Eggum, O. (1989). Nutrient composition and protein composition of minor millets. *Plant Foods for Human Nutrition, 39*, 201–208.

Geetha, C., & Easwaran, P. P. (1990). Hypoglycemic effect of millet incorporated breakfast items on selected non-insulin dependent diabetic patients. *The Indian Journal of Nutrition and Dietetics, 27*(11), 316–320.

Gopalan, C., Rama Sastri, B. V., & Balasubramanian, S. C. (2009). *Nutritive value of Indian foods.* Hyderabad, India: National Institute of Nutrition, Indian Council of Medical Research.

Guha, M., & Malleshi, N. G. (2006). A process for preparation of pre-cooked cereals and vegetables based foods suitable for use as instant soup mix or of similar type foods. Indian Patent 257/NF/06.

Hegde, P. S., Anitha, B., & Chandra, T. S. (2005). *In vivo* effect of whole grain flour of finger millet (*Eleusine coracana*) and kodo millet (*Paspalum scrobiculatum*) on rat dermal wound healing. *Indian Journal of Experimental Biology, 43*(3), 254–258.

Hegde, P. S., & Chandra, T. S. (2005). ESR spectroscopic study reveals higher free radical quenching potential in kodo millet (*Paspalum scrobiculatum*) compared to other millets. *Food Chemistry, 92*, 177–182.

Hegde, P. S., Chandrakasan, G., & Chandra, T. S. (2002). Inhibition of collagen glycation and crosslinking *in vitro* by methanolic extracts of finger millet (*Eleusine coracana*) and kodo millet (*Paspalum scrobiculatum*). *The Journal of Nutritional Biochemistry, 13*, 517–521.

Hegde, P. S., Rajasekaran, N. S., & Chandra, T. S. (2005). Effects of the antioxidant properties of millet species on oxidative stress and glycemic status in alloxan-induced rats. *Nutrition Research, 25*, 1109–1120.

Hulse, J. H., Laing, E. M., & Pearson, O. E. (1980). *Sorghum and the millets: Their composition and nutritive value*. London: Academic Press.

Indian Council of Medical Research (ICMR). (2010). Nutrient requirements and recommended dietary allowances for Indians. A report of the Expert Group of the Indian Council of Medical Research.

Itagi, H. B. N., & Singh, V. (2011). Preparation, nutritional composition, functional properties and antioxidant activities of multigrain composite mixes. *Journal of Food Science and Technology, 49*(1), 74–81.

Itagi, H. B. N., Singh, V., Indiramma, A. R., & Prakash, M. (2011). Shelf stable multigrain *halwa* mixes: Preparation of *halwa*, their textural and sensory studies. *Journal of Food Science and Technology*. http://dx.doi.org/10.1007/s13197-011-0423-z.

Joseph, K., Kurien, P. P., Swaminathan, M., & Subrahmanyan, V. (1959). The metabolism of nitrogen, calcium and phosphorus in under nourished children. *British Journal of Nutrition, 13*, 213–218.

Jung, E. Y., Suh, H. J., Hong, W. S., Kim, D. G., Hong, Y. H., Hong, I. S., et al. (2009). Uncooked rice of relatively low gelatinization degree resulted in lower metabolic glucose and insulin responses compared with cooked rice in female college students. *Nutrition Research, 29*, 457–461.

Jyotsna, R., Soumya, C., Indrani, D., & Venkateswara Rao, G. (2011). Effect of replacement of wheat flour with finger millet flour (*Eleusine corcana*) on the batter microscopy, rheology and quality characteristics of muffins. *Journal of Texture Studies, 42*(6), 478–489.

Kamala, K., Ganesan, A., Shobana, S., Sudha, V., Malleshi, N. G., & Mohan, V. (2010). Dietary management of finger millet. *Current Science (Correspondence), 99*(1), 9.

Kamat, S. S. (2008). A study on documentation and evaluation of indigenous method of preparation of *papad* with special reference to cereals and millets. M.Sc. Dissertation Thesis. University of Agricultural Sciences, Dharward, Karnataka.

Kannan, S. (2010). Finger millet in nutrition transition: An infant weaning food ingredient with chronic disease preventive potential. *British Journal of Nutrition, 104*, 1733–1734.

Kavita, M. S., & Prema, L. (1995). Postprandial blood glucose response to meals containing different carbohydrates in diabetics. *The Indian Journal of Nutrition and Dietetics, 32*, 123–127.

Krishnan, R., Dharmaraj, U., Sai Manohar, R., & Malleshi, N. G. (2011). Quality characteristics of biscuits prepared from finger millet seed coat based composite flour. *Food Chemistry, 129*(2), 499–506.

Krishnan, M., & Prabhasankar, P. (2010). Studies on pasting, microstructure, sensory, and nutritional profile of pasta influenced by sprouted finger millet (*Eleucina coracana*) and green banana (*Musa paradisiaca*) flours. *Journal of Texture Studies, 41*, 825–841.

Kurien, P. P., & Desikachar, H. S. R. (1966). Preparation of refined white flour from ragi using a laboratory mill. *Journal of Food Science and Technology, 3*, 56–58.

Kurien, P. P., Joseph, K., Swaminathan, M., & Subrahmanyan, V. (1959). The distribution of nitrogen, calcium and phosphorus between the husk and endosperm of ragi (*Eleusine coracana*). *Food Science, 8*(10), 353–355.

Kurup, P. G., & Krishnamurthy, S. (1992). Glycemic index of selected food stuffs commonly used in South India. *International Journal for Vitamin and Nutrition Research, 62*, 266–268.

Lakshmi Kumari, P., & Sumathi, S. (2002). Effect of consumption of finger millet on hyperglycemia in non-insulin dependent diabetes mellitus (NIDDM) subjects. *Plant Foods for Human Nutrition, 57*, 205–213.

Lei, V., Friis, H., & Michealsen, K. F. (2006). Spontaneously fermented finger millet product as a natural probiotic treatment for diarrhea in young children: An intervention study in Northern Ghana. *International Journal of Food Microbiology, 110*, 246–253.

Mahadevappa, V. G., & Raina, P. L. (1978). Lipid profile and fatty acid composition of finger millet (*Eleusine coracana*). *Journal of Food Science and Technology, 15*(3), 100–102.

Malathi, D., & Nirmalakumari, A. (2007). Cooking of small millets in Tamil Nadu. In K. T. Krishne Gowda & A. Seetharam (Eds.), *Food uses of small millets and avenues for further processing and value addition* (pp. 57–63): Indian Council of Agricultural Research. Project Coordination cell, All India Co-ordinated small millets improvement project. Indian Council of Agricultural Research, UAS, GKVK, Bangalore, India.

Malleshi, N. G. (2002). Malting of finger millet and specialty foods based on the millet malt. In *Sovenier of XV Indian convention of food scientists and technologists* (pp. 17–23). Mysore: AFST(I).

Malleshi, N. G. (2006). Decorticated finger millet (*Eleusine coracana*) and process for preparation of decorticated finger millet. United States Patent 2006/7029720 B2.

Malleshi, N. G. (2007). Nutritional and technological features of *ragi* (finger millet) and processing for value addition. In K. T. Krishne Gowda & A. Seetharam (Eds.), *Food uses of small millets and avenues for further processing and value addition*; Indian Council of Agricultural Research. Project Coordination cell, All India Co-ordinated small millets improvement project. Indian Council of Agricultural Research, UAS, GKVK, Bangalore, India.

Malleshi, N. G., & Chakravarthy, M. (1994). A process for preparation of food, particularly useful for enteral feeding. 1724/DEL/94.

Malleshi, N. G., & Desikachar, H. S. R. (1982). Formulation of a weaning food with low hot-paste viscosity based on malted ragi (*Eleusine coracana*) and green gram (*Phaseolus radiatus*). *Journal of Food Science and Technology, 19*, 193–197.

Malleshi, N. G., Desikachar, H. S. R., & Tharanathan, R. N. (1986). Free sugars and non-starchy polysaccharides of finger millet (*Eleusine coracana*), pearl millet (*Pennisetum typhoideum*), foxtail millet (*Setaria italica*) and their malts. *Food Chemistry, 20*, 253–261.

Malleshi, N. G., & Gokavi, S. S. (1999). A process for preparation of Infant food. Patent no. 194999.

Mangala, S. L., Malleshi, N. G., Mahadevamma, S., & Tharanathan, R. N. (1999). Resistant starch from differently processed rice and ragi (Finger millet). *European Food Research and Technology, 209*(1), 32–37.

Mangala, S. L., Ramesh, H. P., Udayasankar, K., & Tharanathan, R. N. (1999). Resistant starch derived from processed ragi (finger millet, *Eleusine coracana*) flour: Structural characterization. *Food Chemistry, 64*, 475–479.

Mani, U. V., Prabhu, B. M., Damle, S. S., & Mani, I. (1993). Glycaemic index of some commonly consumed foods in western India. *Asia Pacific Journal of Clinical Nutrition, 2*(3), 111–114.

Matsumoto, N., Ishigaki, F., Ishigaki, A., Iwashina, H., & Hara, Y. (1993). Reduction of blood glucose levels by tea catechin. *Bioscience, Biotechnology, and Biochemistry, 57*, 525–527.

McDonough, C. M., Rooney, L. W., & Earp, C. F. (1986). Structural characteristics of *Eleusine coracana* (Finger millet) using scanning electron and fluorescence microscopy. *Food Microstructure, 5*, 247–256.

Mohan, B. H., Anitha, G., Malleshi, N. G., & Tharanathan, R. N. (2005). Characteristics of native and enzymatically hydrolyzed ragi (*Eleusine coracana*) and rice (*Oryza sativa*) starches. *Carbohydrate Polymers, 59*, 43–50.

Mohan, V., Radhika, G., Sathya, R. M., Tamil, S. R., Ganesan, A., & Sudha, V. (2009). Dietary carbohydrates, glycemic load, food groups and newly detected type 2 diabetes among urban Asian Indian population in Chennai, India (Chennai Urban Rural Epidemiology Study 59). *British Journal of Nutrition, 102*, 1498–1506.

Nadine, N., Nada, A., & Nahla, H. (2004). Glycemic and insulinemic responses to hot vs cooled potato in males with varied insulin sensitivity. *Nutrition Research, 24*(12), 993–1004.

Newby, P. K., Maras, J., Bakun, P., Muller, D., Ferrucci, L., & Tucker, K. L. (2007). Intake of whole grains, refined grains and cereal fiber measured with 7-d diet records and associations with risk of chronic disease. *American Journal of Clinical Nutrition, 86,* 1745–1753.

NNMB. (2006). Diet and nutritional status of rural population and prevalence of hypertension among adults in rural areas. NNMB Technical Report 24. National Institute of Nutrition, Indian Council of Medical Research, Hyderabad, India.

Parada, J., & Aguilera, J. M. (2008). *In vitro* digestibility and glycemic response of potato starch is related to granule size and degree of gelatinization. *Journal of Food Science, 74*(1), E34–E38.

Patel, J. C., Dhirawani, M. K., & Dharne, R. D. (1968). Ragi in the management of diabetes mellitus. *Indian Journal of Medical Sciences, 22,* 28–29.

Pathak, P., Srivastava, S., & Grover, S. (2000). Development of food products based on millets, legumes and fenugreek seeds and their suitability in the diabetic diet. *International Journal of Food Sciences and Nutrition, 51*(5), 409–414.

Pereira, M. A., Jacobs, D. R., Pins, J. J., Raatz, S. K., Gross, M. D., Slavin, J. L., et al. (2002). Effect of whole grains on insulin sensitivity in overweight hyperinsulinemic adults. *American Journal of Clinical Nutrition, 75,* 848–855.

Platel, K., Eipeson, S. W., & Srinivasan, K. (2010). Bioaccessible mineral content of malted finger millet (*Eleusine coracana*), wheat (*Triticum aestivum*), and barley (*Hordeum vulgare*). *Journal of Agricultural and Food Chemistry, 58,* 8100–8103.

Platel, K., & Shurpalekar, K. S. (1994). Resistant starch content of Indian foods. *Plant Foods for Human Nutrition, 45,* 91–95.

Pore, M. S., & Magar, N. G. (1976). Effect of ragi feeding on serum cholesterol level. *Indian Journal of Medical Research, 64*(6), 909–914.

Pradhan, A., Nag, S. K., & Patil, S. K. (2010). Dietary management of finger millet controls diabetes. *Current Science, 98*(6), 763–765.

Radhika, G., Sathya, R. M., Ganesan, A., Saroja, R., Vijayalakshmi, P., Sudha, A., et al. (2011). Dietary profile of urban adult population in south India in the context of chronic disease epidemiology (CURES-68). *Journal of Public Health Nutrition, 14*(4), 591–598.

Radhika, G., Van Dam, R. M., Sudha, V., Ganesan, A., & Mohan, V. (2009). Refined grain consumption and the metabolic syndrome in urban Asian Indians (Chennai Urban Rural Epidemiology Study 57). *Metabolism, Clinical and Experimental, 58,* 675–681.

Raghu, K. S., & Bhattacharya, S. (2010). Finger millet dough treated with α-amylase: Rheological, physicochemical and sensory properties. *Food Research International, 43*(8), 2147–2154.

Rajasekaran, N. S., Nithya, M., Rose, C., & Chandra, T. S. (2004). The effect of finger millet feeding on the early responses during the process of wound healing in diabetic rats. *Biochimica et Biophysica Acta, 1689,* 190–201.

Ramachandra, G., Virupaksha, T. K., & Shadaksharaswamy, M. (1977). Relationship between tannin levels and *in vitro* protein digestibility in finger millet (*Eleusine coracana* Gaertn.). *Journal of Agricultural and Food Chemistry, 25*(5), 1101–1104.

Ramanathan, M. K., & Gopalan, C. (1957). Effect of different cereals on blood sugar levels. *Indian Journal of Medical Research, 45*(2), 255–262.

Rateesh, K., Usha, D., & Malleshi, N. G. (2012). Influence of decortication, popping and malting on bioaccessibility of calcium, iron and zinc in finger millet. *LWT—Food Science and Technology, 48*(2), 169–174.

Ravindran, G. (1991). Studies on millets: Proximate composition, mineral composition, and phytate and oxalate contents. *Food Chemistry, 39,* 99–107.

Rejaul, H. B. (2008). Quality characteristics of finger millet (*Eleusine coracana*) products with special reference to flakes. M.Sc. Dissertation submitted to Tezpur University, Tezpur, Assam.

Roopa, S., & Premavalli, K. S. (2008). Effect of processing on starch fractions in different varieties of finger millet. *Food Chemistry, 106*(3), 875–882.

Roopa, M. R., Urooj, A., & Puttaraj, S. (1998). Rate of *in vitro* starch hydrolysis and digestibility index of ragi based preparations. *Journal of Food Science and Technology, 35*, 138–142.

Saha, S., Gupta, A., Singh, S. R. K., Bharti, N., Singh, K. P., Mahajan, V., et al. (2011). Compositional and varietal influence of finger millet flour on rheological properties of dough and quality of biscuit. *LWT—Food Science and Technology, 44*(3), 616–621.

Sahyoun, N. R., Jacques, P. F., Zhang, X. L., Juan, W., & McKeown, N. M. (2006). Whole-grain intake is inversely associated with the metabolic syndrome and mortality in older adults. *American Journal of Clinical Nutrition, 83*, 124–131.

Sastri, B. N. (1939). Ragi, *Eleucine coracana* Gaertn—A new raw material for the malting industry. *Current Science, 1*, 34–35.

Scalbert, A., Manach, C., Morand, C., Remesy, C., & Jimenez, L. (2005). Dietary polyphenols and the prevention of diseases. *Critical Reviews in Food Science and Nutrition, 45*, 287–306.

Serna-Saldiver, S. O., McDonough, C. M., & Rooney, L. W. (1994). The millets. In K. J. Lorenz & K. Kulp (Eds.), *Hand book of cereal science and technology* (pp. 271–300). New York: Marcel Dekker.

Sharavathy, M. K., Urooj, A., & Puttaraj, S. (2001). Nutritionally important starch fractions in cereal based Indian food preparations. *Food Chemistry, 75*, 241–247.

Shobana, S. (2009). Investigations on the carbohydrate digestibility of finger millet with special reference to the influence of its seed coat constituents. Ph.D. Thesis. University of Mysore, Mysore.

Shobana, S., Harsha, M. R., Platel, K., Srinivasan, K., & Malleshi, N. G. (2010). Amelioration of hyperglycemia and its associated complications by finger millet (*Eleusine coracana* L.) seed coat matter in streptozotocin induced diabetic rats. *British Journal of Nutrition, 104*, 1787–1795.

Shobana, S., & Malleshi, N. G. (2007). Preparation and functional properties of decorticated finger millet (*Eleusine coracana*). *Journal of Food Engineering, 79*, 529–538.

Shobana, S., Sreerama, Y. N., & Malleshi, N. G. (2009). Composition and enzyme inhibitory properties of finger millet (*Eleusine coracana* L.) seed coat phenolics: Mode of inhibition of α-glucosidase and α-amylase. *Food Chemistry, 115*, 1268–1273.

Shobana, S., Ushakumari, S. R., Malleshi, N. G., & Ali, S. Z. (2007). Glycemic response of rice, wheat and fingermillet based diabetic food formulations in normoglycemic subjects. *International Journal of Food Sciences and Nutrition, 58*(5), 363–372.

Shukla, K., & Srivastava, S. (2011). Evaluation of finger millet incorporated noodles for nutritive value and glycemic index. *Journal of Food Science and Technology*. http://dx.doi.org/10.1007/s13197-011-0530-x.

Singh, P. (2008). History of millet cultivation in India. In L. Gopal & V. C. Srivastava (Eds.), *History of science, philosophy and culture in Indian civilisation, Volume V, Part I, History of Agriculture in India (up to c.1200 AD)* (pp. 107–119). New Delhi: PHISPC, Centre for Studies in Civilisations. Concept Publishing Company.

Singh, K. P., Abhinav, M., & Mishra, H. N. (2012). Fuzzy analysis of sensory attributes of bread prepared from millet-based composite flours. *LWT—Food Science and Technology, 48*(2), 276–282.

Singh, V., & Ali, S. Z. (2006). *In vitro* hydrolysis of starches by α-amylase in comparison to that by acid. *American Journal of Food Technology, 1*(1), 43–51.

Sripriya, G., Antony, U., & Chandra, T. S. (1997). Changes in carbohydrate, free aminoacids, organic acids, phytate and HCL extractability of minerals during germination and fermentation of finger millet (*Eleusine coracana*). *Food Chemistry, 58*(4), 345–350.

Sripriya, G., Chandrasekharan, K., Murthy, V. S., & Chandra, T. S. (1996). ESR spectroscopic studies on free radical quenching action of finger millet (*Eleusine coracana*). *Food Chemistry, 57*(4), 537–540.

Stephen, M. M., John, V. C., Peter, R. S. M., Wilfried, O., Patrick, K., & Andre, H. (2002). Evaluation of the nutritional characteristics of a finger millet based complementary food. *Journal of Agricultural and Food Chemistry, 50*, 3030–3036.

Subba Rao, M. V. S. S. T., & Muralikrishna, G. (2002). Evaluation of the antioxidant properties of free and bound phenolic acids from native and malted finger millet (Ragi, *Eleusine coracana* Indaf-15). *Journal of Agricultural and Food Chemistry, 50*, 889–892.

Subrahmanyan, V., Narayana Rao, M., Rama Rao, G., & Swaminathan, M. (1955). The metabolism of nitrogen, calcium and phosphorus in human adults on a poor vegetarian diet containing ragi (*Eleusine coracana*). *British Journal of Nutrition, 9*, 350–357.

Sudha, M. L., Vetrimani, R., & Rahim, A. (1998). Quality of vermicelli from finger millet (*Eleusine coracana*) and its blend with different milled wheat fractions. *Food Research International, 31*(2), 99–104.

Sumathi, A., Vishwanatha, S., Malleshi, N. G., & Venkat Rao, S. (1997). Glycemic response to malted, popped and roller dried wheat-legume based foods in normal subjects. *International Journal of Food Sciences and Nutrition, 48*, 103–107.

Tatala, S., Ndossi, G., Ash, D., & Mamiro, P. (2007). Effect of germination of finger millet on nutritional value of foods and effect of food supplement on nutrition and anaemia status in Tanzanian children. *Tanzania Health Research Bulletin, 9*(2), 77–86.

Thapa, S., & Tamang, J. P. (2004). Product characterization of kodo ko jaanr: Fermented finger millet beverage of the Himalayas. *Food Microbiology, 21*(5), 617–622.

Tovey, F. I., Jayaraj, A. P., & Clark, C. G. (1975). The possibility of dietary protective factors in duodenal ulcer. *Postgraduate Medical Journal, 51*, 366–372.

Tripathi, B., & Platel, K. (2010). Finger millet (*Eleucine coracana*) flour as a vehicle for fortification with zinc. *Journal of Trace Elements in Medicine and Biology, 24*(1), 46–51.

Tripathi, B., & Platel, K. (2011). Iron fortification of finger millet (*Eleucine coracana*) flour with EDTA and folic acid as co-fortificants. *Food Chemistry, 126*(2), 537–542.

Tripathi, B., Platel, K., & Srinivasan, K. (2012). Double fortification of sorghum (*Sorghum bicolor* L. Moench) and finger millet (*Eleucine coracana* L. Gaertn) flours with iron and zinc. *Journal of Cereal Science, 55*(2), 195–201.

Ugare, R., Chimmad, B., Naik, R., Bharati, P., & Itagi, S. (2011). Glycemic index and significance of barnyard millet (*Echinochloa frumentacae*) in type II diabetics. *Journal of Food Science and Technology.* http://dx.doi.org/10.1007/s13197-011-0516-8.

Umapathy, P. K., & Kulsum, A. (1976). Ragi—A poor man's millet. *Journal of the Mysore University Section A, XXXVII*, 45–48.

Urooj, A., & Puttaraj, S. (1999). Digestibility index and factors affecting rate of starch digestion *in vitro* in conventional food preparation. *Die Nahrung, 43*, 42–47.

Urooj, A., Rupashri, K., & Puttaraj, S. (2006). Glycemic responses to finger millet based Indian preparations in non-insulin dependent diabetic and healthy subjects. *Journal of Food Science and Technology, 43*(6), 620–625.

Ushakumari, S. R. (2009). Technological and physico-chemical characteristics of hydrothermally-treated finger millet. Ph.D. Thesis, University of Mysore, Mysore.

Ushakumari, S. R., Rastogi, N. K., & Malleshi, N. G. (2007). Optimization of process variables for the preparation of expanded finger millet using response surface methodology. *Journal of Food Engineering, 82*(1), 35–42.

Varsha, V., Urooj, A., & Malleshi, N. G. (2009). Evaluation of antioxidant and antimicrobial properties of finger millet (*Eleusine coracana*) polyphenols. *Food Chemistry, 114*(1), 340–346.

Veenashri, B. R., & Muralikrishna, G. (2011). *In vitro* anti-oxidant activity of xylo-oligosaccharides derived from cereal and millet brans—A comparative study. *Food Chemistry, 126*, 1475–1481.

Venkateswaran, V., & Vijayalakshmi, G. (2010). Finger millet (*Eleusine coracana*)—An economically viable source for antihypercholesterolemic metabolites production by *Monascus purpureus*. *Journal of Food Science and Technology, 47*(4), 426–431.

Venn, B. J., & Green, T. J. (2007). Glycemic index and glycemic load: Measurement issues and their effect on diet–disease relationships. *European Journal of Clinical Nutrition, 61* (Suppl 1), S122–S131.

Vidyavati, H. G., Mushtari, J., Vijayakumari, J., Gokavi, S. S., & Begum, S. (2004). Utilization of finger millet in the preparation of *papad*. *Journal of Food Science and Technology, 41*, 379–382.

Vijayakumar, P., Mohankumar, J. B., & Srinivasan, T. (2010). Quality evaluation of noodles from millet flour blend incorporated composite flour. *Electronic Journal of Environmental, Agricultural and Food Chemistry, 9*(3), 479–492.

Virupaksha, T. K., Ramachandra, G., & Nagaraju, D. (1975). Seed proteins of finger millet and their amino acid composition. *Journal of the Science of Food and Agriculture, 26*, 1237–1246.

Wankhede, D. B., Shehnaj, A., & Raghavendra Rao, M. R. (1979a). Carbohydrate composition of finger millet (*Eleusine coracana*) and foxtail millet (*Setaria italica*). *Qualitas Plantarum Plant Foods for Human Nutrition, 28*, 293–303.

Wankhede, D. B., Shehnaj, A., & Raghavendra Rao, M. R. (1979b). Preparation and physicochemical properties of starches and their fractions from finger millet (*Eleusine coracana*) and foxtail millet (*Setaria italica*). *Starch, 31*, 153–159.

Wolever, T. M. S., Jenkins, D. J. A., Jenkins, A. L., & Joss, R. G. (1991). The glycemic index: Methodology and clinical implications. *American Journal of Clinical Nutrition, 54*, 846–854.

CHAPTER TWO

Advances in Food Powder Agglomeration Engineering

B. Cuq[*,†,‡,1], C. Gaiani[§], C. Turchiuli[¶], L. Galet[‖], J. Scher[§], R. Jeantet[#,**],
S. Mandato[*,†,‡], J. Petit[§,#,**,‡‡], I. Murrieta-Pazos[§,‖], A. Barkouti[‡,¶],
P. Schuck[#,**], E. Rondet[††], M. Delalonde[††], E. Dumoulin[¶],
G. Delaplace[‡‡], T. Ruiz[*,†,‡]

[*]Montpellier SupAgro, UMR 1208 IATE, Montpellier, France
[†]INRA, UMR 1208 IATE, Montpellier, France
[‡]UMR 1208 IATE, Université Montpellier 2, Montpellier, France
[§]LiBio, Université de Lorraine, Nancy, France
[¶]AgroParisTech, UMR GENIAL, Massy, France
[‖]Ecole des Mines Albi, UMR RAPSODEE, Albi, France
[#]AgroCampusOuest, UMR 1253 STLO, Rennes, France
[**]INRA, UMR 1253 STLO, Rennes, France
[††]UMR QualiSud, Université Montpellier 1, Montpellier, France
[‡‡]INRA, UR 638 PIHM, Villeneuve d'Ascq, France
[1]Corresponding author: e-mail address: cuq@supagro.inra.fr

Contents

Advances in Food and Nutrition Research, Volume 69
ISSN 1043-4526
http://dx.doi.org/10.1016/B978-0-12-410540-9.00002-8

Abstract

Food powders are used in everyday life in many ways and offer technological solutions to the problem of food production. The natural origin of food powders, diversity in their chemical composition, variability of the raw materials, heterogeneity of the native structures, and physicochemical reactivity under hydrothermal stresses contribute to the complexity in their behavior. Food powder agglomeration has recently been considered according to a multiscale approach, which is followed in the chapter layout: (i) at the particle scale, by a presentation of particle properties and surface reactivity in connection with the agglomeration mechanisms, (ii) at the mechanisms scale, by describing the structuration dynamics of agglomerates, (iii) at the process scale, by a presentation of agglomeration technologies and sensors and by studying the stress transmission mode in the powder bed, and finally (iv) by an integration of the acquired knowledge, thanks to a dimensional analysis carried out at each scale.

1. INTRODUCTION

Since the 1980s, industrialists, equipment suppliers, and scientists have been aware of the relevance of a global powder-engineering approach to the stakes in the production and use of powders. Even if the generic degree still remains to be defined, it is clear that the science of granular matter is gradually growing. The relevance is based on the integration of disciplines related to the application domains (food science and technology) with academic disciplines (such as process engineering, physicochemistry, and physics). The understanding of the behavior of the granular matter constituted by dense assemblies of solid grains, under mechanical, hydric, thermal, or chemical stresses, remains a relevant issue.

Food powders are used in everyday life for many purposes (e.g., salt, pepper, spices, sugar, flour, coffee, almond powder, coloring agents, etc.) and offer technological solutions to face the complexity of food production. Food powders are easy to preserve, transport, store, and process. The natural origin of food powders, the diversity of the chemical composition, variability of the raw materials, heterogeneity of the native structures, and physicochemical reactivity under hydrothermal stresses contribute to the complexity in their behavior. Contrary to model powders, which are made by homogeneous, spherical, monodisperse, and inert particles (e.g., glass beads), food powders display great heterogeneity in size, shape, and structure. Food powders are reactive material when wetted or submitted to high temperatures, as their molecules undergo irreversible physicochemical changes. The reactivity finds its origin at different scales: the molecular, supramolecular, and microscopic scales. Although significant scientific work has been conducted over the last 15 years (Barbosa-Canovas, Ortega-Rivas, Juliano, & Yan, 2005; Cuq, Rondet, & Abecassis, 2011; Fitzpatrick & Ahrné, 2005), descriptions of food powders still remain incomplete.

Agglomeration is an operation during which particles are assembled to form bigger agglomerates, in which the original particle can still be distinguished. It is implemented for a great variety of powders (mineral, organic, biological, amorphous, crystalline, spherical, smooth, irregular, etc.). It improves the powders' functionalities (e.g., flow properties, dust generation, explosion risks, storage, mixing capacity, aspect, dispersion, solubility, and controlled release) (Iveson, Litster, Hapgood, & Ennis, 2001; Litster & Ennis, 2004; Palzer, 2011; Saleh & Guigon, 2009). The process is used to agglomerate food ingredients, for example, crystalline sucrose, fibers, seasoning cubes, instant tea, aspartame, icing sugar, yeast extracts, corn syrups, whey powders, couscous grain, protein powders, bakery mixes, and flavor powders (Palzer, 2011; Saad, Barkouti, Rondet, Ruiz, & Cuq, 2011; Yusof, Smith, & Briscoe, 2005). While the industrial applications are widespread, the scientific description of the agglomeration of food powders is much more recent and remains incomplete (Palzer, 2011).

Due to the multiplicity of mechanisms, the theoretical aspects of agglomeration are not yet fully understood to allow predicting quantitatively the effect of the physicochemical properties and the operating conditions on the quality of agglomerates, growth rate, and main mechanisms. Besides, because most of the scientific studies have been carried out with chemically inert particles (e.g., glass beads), the applicability of the results to food agglomeration processes is not directly possible. The food industry still lacks theoretical foundations, which is a prerequisite for improving the control and efficiency of industrial processes. The agglomeration of food powders requires taking into consideration the physicochemical phenomena, based on irreversible changes of particles, and the contribution of viscous forces to stabilize the agglomerates. The introduction of more fundamental

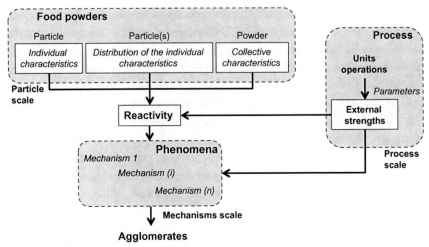

Figure 2.1 Multiscale approach to food powder agglomeration.

sciences, such as the physics of soft matter or the mechanics of granular matter, coupled with modern techniques of investigation, would allow identifying the mechanisms necessary for the process parameters. These challenges are complex and must be managed at once and in a coherent way. The studies have also to integrate the complexity of food powders as their physicochemical reactivity at different scales is not yet fully understood. To specify the agglomeration theories, the efforts at understanding must lead to answering two fundamental questions: How do particles join? How do particles move?

Food powder agglomeration has recently been considered on a multiscale approach, which constitutes the chapter layout (Fig. 2.1): (i) at the particle scale, by considering the particle properties and surface reactivity in connection with the agglomeration mechanisms, (ii) at the mechanisms scale, by describing the structuration dynamics of agglomerates, (iii) at the process scale, by a presentation of agglomeration technologies and sensors and by studying the stress transmission mode in the powder bed, and finally (iv) by an integration approach, thanks to dimensional analysis (DA) methods.

2. FOOD POWDER REACTIVITY AND SURFACE PROPERTIES

The agglomeration of food powders demands the consideration of the specific phenomena associated with their physicochemical reactivity. Food powders are reactive powders because they are able to undergo partially irreversible changes when subjected to external stresses, such as hydration, heating, and/or strains. The reactivity finds its origin at different scales: molecular, supramolecular, and microscopic. A global process—a product approach—is considered to integrate the physicochemical reactivity of the food powders as a key contributor of the agglomeration mechanisms. At the molecular scale, food powders can be classified by the nature of their constitutive molecules, as organic molecules or minerals. The physicochemical reactivity at the molecular scale is associated with polarity and chemical interactions. The polarity properties of food powders affect the capacity of molecules to interact with the binder liquids that are used for agglomeration and thus contribute to their technological behavior. Naturally, the polarities of the wetting liquid and particles have to be compatible. Based on their chemical composition, food molecules have the capacity to establish, between them and with the surrounding molecules, interactions of different energies. The physicochemical reactivity of food powders is also defined at the supramolecular scale, by considering the degree of organization of the molecules and macromolecules and the plasticization mechanisms. At the particle scale, surface polarity, water absorption capacity, surface solubility of molecules, and surface stickiness have to be taken into consideration.

An important problem facing researchers and manufacturers is the lack of a central source of information that provides practical knowledge focused on food powder reactivity and surface properties. New and pioneering methodologies have recently been considered to determine these factors.

2.1. Importance of particle surface reactivity

The relationship between the functional properties of the particles (e.g., flowability, wettability, and hydration) and their surface properties is a key factor and a challenge in food science developments. A large number of raw ingredients from various sources (e.g., animals, vegetables, and inorganics) in many food industries are submitted to processes to obtain a specific dry form such as comminution, agglomeration, spray drying, dry mixing, coating, and encapsulation. (Cuq et al., 2011). All these processes, as also the packaging, storage, and transport conditions, may influence the powder surface composition and the surface reactivity (Fig. 2.2). It is now generally well accepted that surface composition has a strong impact on many powder functional properties, for instance, rehydration (Forny, Marabi, & Palzer, 2011; Gaiani et al., 2006), hydration (Murrieta-Pazos et al., 2011), solubility (Fyfe, Kravchuk, Le, et al., 2011; Haque, Bhandari, Gidley, Deeth, & Whittaker, 2011; Jayasundera, Adhikari, Howes, & Aldred, 2011; McKenna, Lloyd, Munro, & Singh, 1999), caking (Hartmann & Palzer, 2011; Montes, Santamaria, Gumy, & Marchal, 2011;

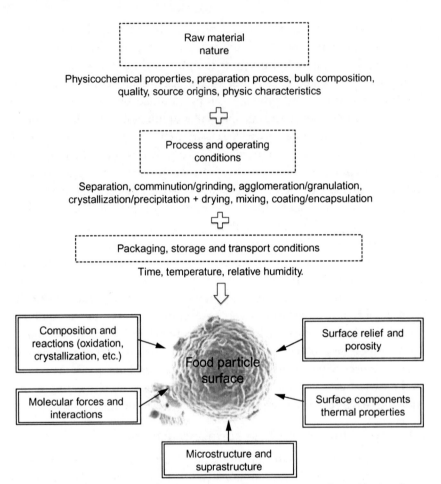

Figure 2.2 Factors affecting food powder surface properties. *From Murrieta-Pazos, Gaiani, Galet, Calvet, et al. (2012).* (For color version of this figure, the reader is referred to the online version of this chapter.)

Prime, Leaper, et al., 2011; Prime, Stapley, Rielly, Jones, & Leaper, 2011), flowability (Elversson & Millqvist-Fureby, 2006; Kim, Chen, & Pearce, 2005a; Montes, Santamaria, et al., 2011; Szulc & Lenart, 2010), and sticking (Bunker, Zhang, Blanchard, & Roberts, 2011; Chen & Ozkan, 2007; Chuy & Labuza, 1994; Ozkan, Walisinghe, & Chen, 2002). etc.

2.2. New methodologies allowing surface characterization

Recently, several methodologies have been developed to better characterize food powder surfaces (e.g., composition, energy, structure, shape, and roughness). A recent review highlights some techniques that enable the characterization of food powder surfaces and tries to connect this information with their functional properties (Murrieta-Pazos, Gaiani, Galet, Calvet, et al., 2012). These techniques can be categorized into four groups:

i. The most important developments in microscopy are the increasing use of atomic force microscopy and confocal laser scanning microscopy in the analysis of powder surfaces. It is likely that these two techniques will complement light and electron microscopy methods and play an important role in food surface analysis in the future.

ii. Recent developments in spectroscopy techniques that have been reported are X-ray photoelectron spectroscopy (Gaiani et al., 2007; Jafari, He, & Bhandari, 2007; Rouxhet et al., 2008; Saad, Gaiani, Mullet, Scher, & Cuq, 2011) and environmental scanning electron microscopy, coupled with electron diffraction X-ray (Murrieta-Pazos, Gaiani, Galet, & Scher, 2012). The first one was used to determine "top" surface (around 5 nm) atomic composition and the second one to determine deeper (around 1 μm) surface atomic composition. With the combination of these two techniques, gradients from the surface to the core of whole and skim milk powders can be described (Murrieta-Pazos, Gaiani, Galet, & Scher, 2012).

iii. Surface sorption techniques have also been used to obtain surface information. From the sorption isotherms obtained with dynamic vapor sorption equipment, it is possible to obtain valuable information on different aspects: thermodynamic (enthalpies of sorption and desorption, water activity, crystallization, etc.), structural (specific area, pore size and volume, amorphous state, etc.), and technological (drying condition, stability, handling, storage, packaging, etc.) (Gaiani et al., 2009; Murrieta-Pazos, Gaiani, Galet, Calvet, et al., 2012; Saad et al., 2009). For example, inverse gas chromatography can provide different physical chemistry parameters of the solid surface (surface energy, thermal transitions, crystallinity, specific surface area, etc.) (Conder & Young, 1979).

iv. Surface extraction techniques have also been used to extract some components (e.g., fat) from the powder surface. The extracted fractions can then be analyzed by different techniques, such as differential scanning calorimtry (DSC), gas chromatography (GC), or high performance liquid chromatigraphy (HPLC) (Kim, Chen, & Pearce, 2005b; Murrieta-Pazos, Gaiani, Galet, Calvet, et al., 2012; Vignolles, Lopez, Ehrhardt, et al., 2009; Vignolles, Lopez, Madec, et al., 2009). For instance, a segregation of milk fat within the powders, linked to the fusion temperature and the composition (long/short chain) of these fats, has been demonstrated.

It is worth noting that some techniques still under development may be promising for the characterization of powder surfaces. For example, NMR microimaging experiments were used in the medical field to follow water penetration in powders and the associated drug release kinetics (Dahlberg, Millqvist-Fureby, Schuleit, & Furo, 2010). The same technique was also applied with success to study water transport in porous silica (Aristov et al., 2002) and water distribution during the drying process of gelatin gel (Ruiz-Cabrera, Foucat, Bonny, Renou, & Daudin, 2005). Another nondestructive tool has been employed to determine the uniformity and the repartition of a drug compound in an inhalable powder by confocal Raman microscopy (Schoenherr, Haefele, Paulus, & Francese, 2009). The potentiality of X-ray microtomography has been investigated (Perfetti, Van de Casteele, Rieger, Wildeboer, & Meesters, 2009) to obtain morphological and surface characterization of dry particles and, in particular, of their surface layer. Indeed, a high level of details at both the micro and macroscale could be obtained, such as density, porosity, surface/volume ratio, or thickness of the coating layer. Another tomographic technique, the tomographic atom probe, was used to provide, at the atomic scale, the spatial distribution of atoms in the analyzed specimen (Larde, Bran, Jean, & Le Breton, 2009).

2.3. Factors affecting powder surfaces and reactivity

In parallel to these investigations of food powder surfaces, the evolution of the surface in response to external factors is of great interest in understanding and optimizing food powder processing, in particular, during agglomeration. From the generation process (milling, agglomeration, and storage) to end-use utilization (e.g., wetting, rehydration, dispersion), many factors are integrated. Indeed, food particles are particularly sensible to their surroundings (e.g., temperature and relative humidity) and have a great propensity to react and to participate in reactions, such as surface solubility, caking, and crystallization, and enhance some irreversible reactions and transformations. This response to factors can be summarized by the "reactivity" term (Fig. 2.3). The "surface reactivity" study is of great interest in the development of food powder reactions to process steps. The surface reactivity has to be taken into account in a multiscale approach, from the individual particle to the bulk powder behavior and from an atomic or chemical group scale to a molecular and general surface modification stage.

When considering the spray-drying process, many of the powder surface properties are known to be determined to a great extent by the characteristics and composition of the liquid feed. For instance, the presence of

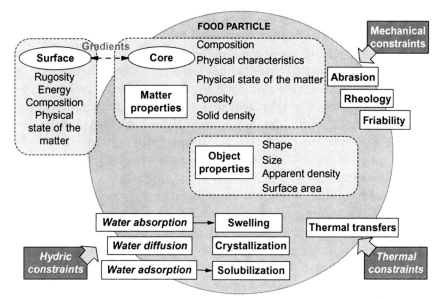

Figure 2.3 Links between characterization, physicochemistry process, and constraints (water, thermal, and mechanical). *From Murrieta-Pazos, Gaiani, Galet, Calvet, et al. (2012).* (For color version of this figure, the reader is referred to the online version of this chapter.)

crystalline substances in the liquid was found to decrease the crystallinity of the final product (Palzer, Dubois, & Gianfrancesco, 2012). The feed solid content is also a key parameter. At high feed solid contents, less fat and protein (i.e., more lactose) appear on the surface of dairy powders (Kim, Chen, & Pearce, 2003, 2009a, 2009b). For cocoa powders, increasing the amounts of carbohydrate before drying could negatively affect the storage stability of the final powder (Montes, Dogan, et al., 2011).

The processes involved in creating the powder properties need close attention (Palzer et al., 2012; Palzer & Fowler, 2010). The reactivity of industrial spray-dried powders was found to be strongly determined by the spray-drying process and conditions employed (drying temperature and degree of homogenization). During drying, a great number of chemical reactions are accelerated, principally due to the use of elevated temperatures. Consequently, the powder structure could be changed mainly during the following steps: concentration, drying (temperature), and homogenization (Palzer et al., 2012). Maillard reactions are accelerated during drying, and various components (aldehydes, aromatic substances, CO_2, etc.) are released. The spray-drying temperature may influence the surface composition of dairy powders

(Gaiani et al., 2010; Kim et al., 2009a, 2009b). The number of homogenization passes reduce the fat globule size, and consequently the amount of fat present on the powder surface. The bigger fat droplets migrate easily and quickly to the surface prior to the surface formation or are present at the surface of droplets leaving the atomizer (Kim et al., 2009b; Vignolles, Lopez, Madec, et al., 2009). The dryer type (industrial-, pilot- and laboratory-scale dryers) significantly influences the powder surfaces. For example, milk powder produced by small-scale dryers does not accurately represent the powder produced by industrial dryers, both in surface composition and morphology (Fyfe, Kravchuk, Nguyen, Deeth, & Bhandari, 2011).

A good knowledge of the processes influencing the powder properties by degrading them (e.g., handling, transport, and storage) is imperative (Fitzpatrick & Ahrné, 2005). Water activity (a_w) and glass transition (T_g) concepts are two theoretical bases developed to determine food powder functionalities (Montes, Dogan, et al., 2011; Rahman, 2012). A large number of surface modifications were observed during storage, such as fat migration at the surface of dairy powders under humid conditions (Faldt & Bergenstahl, 1996) and lactose crystallization at the powder surface.

In this context, the surface characterization of food powders, the relation to the generation process conditions, and their contribution to the transformation mechanisms have become a major research domain. Taking into account the surface reactivity and the end-use properties of food powders, food science has a large research field to explore.

3. HYDROTEXTURAL DIAGRAM

During wet granulation, granular media can be composed of two or three phases, that is, composed of solid and air (S/G), solid and water (S/L), or the three phases of solid/liquid/gas (S/L/G). Depending on the relative amounts of each phase, the medium may take many forms and specific textures. The texture describes the granular layout, that is, compactness (ratio of solid grain volume and the total volume). Shape and texture are the essential characteristics of the state of a granular medium; they are induced by the processing path followed by the granular medium during its technological history. At the scale of a representative elementary volume (REV), multiple physical states can be taken by the granular media and it seems interesting to represent them by a similar description. For a representative scale, humid granular medium is considered a nonsaturated heterogeneous medium constituted by the superposition of three different dispersed phases: solid (native particles), liquid (aqueous solution), and gas (mixture of air and water vapor). For this triphasic thermodynamic system, the definitions of the porous media science (Coussy, 2011) and soil mechanics (Gudehus, 2011; Marshall, Holmes, & Rose, 1996) are adopted. When each phase is deformable, the knowledge of six values (three masses and three volumes) is necessary to realize the material balances of the three phases (Fig. 2.4).

True and apparent densities are defined from these six values (Table 2.1). These values are measured from the determination of solid true densities (pycnometer methods), liquid density, and two masses or two volumes. These values allow the calculation of the standard variables, some of which are defined as follows: solid true density $d_s^* = \rho_s^*/\rho_w^*$, mass water content $w = m_w/m_s$, liquid saturation degree $S_w = \theta_w/(1 - \theta_s)$, and gas saturation degree $S_g = \theta_g/(1 - \theta_s)$ with $S_e + S_g = 1$, compactness $\phi = \theta_s$, porosity $n = 1 - \phi = \theta_w + \theta_g$, and void index $e = (1 - \phi)/\phi$. For wet granular media, the following assumptions may be made: (i) Particles are considered to be intrinsically nondeformable with regard to the deformation of the whole. This study is limited to water granular media states for which the contact between the solid particles is maintained. This condition excludes fluidized media and soft suspensions; (ii) the liquid phase is noncompressible; (iii) the gaseous phase is in thermodynamic equilibrium with the liquid phase (air dissolution and water vaporization are excluded).

The variation in the whole volume is due to grain rearrangement and involves only a change in gas volume. So, the gas volume fraction (θ_g) is the quantity to follow and we propose to follow it through the relationship between the saturation degree and the water content (Eq. 2.1):

$$\phi^{-1} = 1 + d_s^* \frac{w}{S} \qquad [2.1]$$

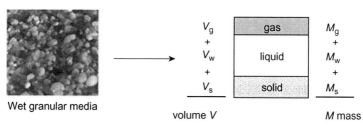

Wet granular media

volume V   M mass

Figure 2.4 Schematization of the real porous medium and definition of the extensive quantities attached to the volume and mass balances. Subscripts g, w, and s are respectively related to gaseous, liquid, and solid phases. (For color version of this figure, the reader is referred to the online version of this chapter.)

Table 2.1 Definition of the descriptive parameters in relation to the present masses and volumes

True density	Apparent density	Volume fractions
$\rho_s^* = \frac{m_s}{V_s}$	$\rho_s = \frac{m_s}{V}$	$\theta_s = \frac{V_s}{V} = \frac{\rho_s}{\rho_s^*}$
$\rho_w^* = \frac{m_w}{V_w}$	$\rho_w = \frac{m_w}{V}$	$\theta_w = \frac{V_w}{V} = \frac{\rho_w}{\rho_w^*}$
$\rho_g^* = \frac{m_g}{V_g}$	$\rho_g = \frac{m_g}{V}$	$\theta_g = \frac{V_g}{V} = \frac{\rho_g}{\rho_g^*}$
$\rho^* = \theta_s\rho_s^* + \theta_w w_s^* + \theta_g\rho_g^*$	$\rho = \rho_s + \rho_w + \rho_g$	$1 = \theta_s + \theta_w + \theta_g$

Considering that the true densities of solid particles ρ_s^* and wetting liquid ρ_w^* are known (values experimentally measured and dependent on temperature, pressure, etc.) and ignoring the gaseous phase mass in relation to the other two phases ($M_g \ll M_s$, M_w), three nonredundant variables provide a complete description of the extensive aspects of the system (V, M_w, M_s). These quantities can be measured by specific experimental trials. The liquid and solid can be respectively replaced by (i) volume or mass water content M_w, and (ii) dry apparent density or porosity or void ratio or compactness for M_s. At the REV scale, two variables are sufficient: (ϕ, w) or (n, S), for example. We can notice that when the medium is rigid (V = constant) or saturated by the liquid, a single value is sufficient.

This framework can integrate the particles with intragranular porosity (bed agglomerates) usually found in food powders. In this situation, a double porosity appears, that of "small pore diameter" (in the agglomerates) and that of "large pore diameter", between the agglomerates. This last point allows the representation of the same form as the evolution of a granular medium. During the wetting, the medium agglomerates through intermediate structures such as the (cf. Section 4) nuclei, agglomerates, and pieces of dough (Fig. 2.5).

3.1. Concept of hydrotextural states

In agreement with earlier Matyas and Radhakrishna (1968) and Liu and Nagel (1998) work, the phase diagram of a wet granular medium is a graph drawn with the following coordinates: compactness (ϕ), water content (w), and the loading parameter (Σ) related to the process considered. When the medium is unloaded, we observe in the (w, ϕ) plane, some characteristic hydrotextural states in relation to saturation of the voids, consistence state, connected space of each phase (relative quantities in relation to the existence of a path-connected space in the topological sense), and thermodynamic states of the fluids. The hydrotextural diagram is the map plotted in the

Figure 2.5 Pictures of nuclei, agglomerate, and paste states (case of kaolin). (For color version of this figure, the reader is referred to the online version of this chapter.)

(w, ϕ) plane, that is, $\Sigma \equiv 0$, which contains these characteristics. It represents the "map of reference" of a phase diagram (w, ϕ, Σ).

3.2. Saturated states

Saturated states of granular media are easiest to describe. With respect to liquid incompressibility hypothesis, the whole volume is equivalent to the sum of solid and liquid volumes. This sum leads to a relation between water content and compactness (Eq. 2.2 with saturation degree equal to 1):

$$\phi_{sat} = \frac{1}{1 + d_s^* w} \qquad [2.2]$$

The ratio of particles and liquid densities (d_s^*) is the only parameter that controls the saturation state. The saturation curve is the first information that appears on the hydrotextural diagram (Fig. 2.6). The hydrotextural area below the saturation curve brings together the unsaturated states. Above this curve, the media are no longer a mixture. The liquid and solid phases are separated (liquid exudation, sedimentation). In some cases, it is necessary to analyze d_s^* with the water content or with the process parameters. This is the case when the solid could have a change in density with wetting (hydration), for example. For these situations, a mass variation arrives with a volume variation. For example, for starch and water mixtures, the true density is not constant (Roman-Gutiérrrez, 2002). This is due to the grain swelling associated with the water uptake. If the hydrated grains are water saturated, it is possible to establish a relationship between the true density and the granular medium water content: $d_s^*(w)$, given by Eq. (2.3):

$$d_s^*(w) = \left(d_{smax}^* - d_{smin}^*\right) e^{-(w/w^*)} + d_{smin}^* \qquad [2.3]$$

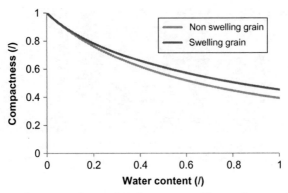

Figure 2.6 Saturation curve. Comparison between swelling and nonswelling hydrated starches ($d^*_{smax} = 1.55$; $d^*_{smin} = 1.2$; $w^* = 30\%$). (For color version of this figure, the reader is referred to the online version of this chapter.)

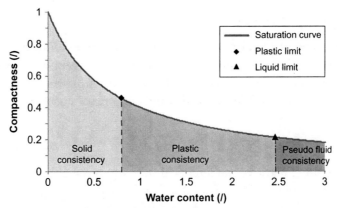

Figure 2.7 Example of plasticity and liquidity limits of microcrystalline cellulose on the hydrotextural diagram. (For color version of this figure, the reader is referred to the online version of this chapter.)

where d^*_{smax} is the maximum true density of nonhydrated grains, d^*_{smin} is the minimum true density of hydrated grains (describing the maximum swelling), and w^* is a parameter linked to the swelling and homogeneous with a water content. Figure 2.6 provides an illustration of the difference between the saturation curves of swelling and nonswelling media.

3.3. Rheological states

The rheological consistency of a wetted granular medium completely changes with the water content (Fig. 2.7). Plastic (w_P) and liquid (w_L) limits described in geotechnics by Atterberg (Atterberg, 1911; Marshall et al., 1996) are the threshold of consistency change. Although their experimental determination is rather simple, they are reliable and reproducible criteria. It is important to note that literature provides little information concerning the textural states (compactness) of the samples for which the Atterberg limits are defined. The results of Rondet, Ruiz, Delalonde, and Desfours (2009) on microcrystalline cellulose, calcium phosphate, and kaolin are in agreement with those of Jefferson and Rogers (1998) showing that these limits are obtained for almost-saturated states. The experimental Atterberg limits remain true only for granular media close to a saturation state.

3.4. Path-connected space of the solid phase

The entire zone below the saturation curve is not "attainable" by granular media. Some kinds of packing are experimentally impossible to achieve. When the particles are very close, the maximal possible compactness is reached. For example, the jamming compactness threshold of the compact packing of monodisperse spheres is close to 0.74. In contrast, the compactness reaches a minimum value when the granular contacts network is less.

In the case of granular media packing, the compactness value is between two limit values, the random loose packing (RLC) and the random close packing (RCP). The RLC is equivalent to the "fluidization" limit, when the contact network between grains is broken. Consequently, it is the limit of solid-phase path-connected space. Inversely, the compactness increase is limited by the jamming threshold (RCP). Its value depends on granular media characteristics (particle shape, size, disparity size, value) and on surface reactivity. It appears experimentally that jamming compactness depends on the technological mode and on process parameters. Depending on the densification process (static and dynamic compaction, drying, vibrations), the jamming compactness value differs greatly.

3.5. Path-connected space of the fluid phases

Sorption or imbibition/drainage tests allow the identification of hygroscopic, pendular, and funicular water states. The sorption isotherms make it possible to define the water content corresponding to the saturation of the granular-specific surface by water molecules (w_{mono}) and characterizing the entry in the pendular field (w_{pendul}), which limits the hygroscopic field and entry into the capillary regime. Beyond this limit, the system has enough interfacial energy to ensure the establishment of capillary bridges between solid particles and above the linked water layers, that is, the Laplace pressure reaches the order of magnitude of the disjunction pressure (Derjaguin, 1989). The modeling of the isotherm, thanks to the Guggenheim–Anderson–Boer equation (Le Meste, Lorient, & Simatos, 2002), allows the identification of the two previous limits. The compactness values corresponding to these specific water contents are recorded. By plotting the values (w_{mono}, ϕ_{mono}) and (w_{pendul}, ϕ_{pendul}) on the hydrotextural diagram, hydrotextural zones of water-adsorbed monolayer and of meniscus appearance are defined. These zones show the end of the hygroscopic domain and entry into the pendular domain.

The capillary retention curves used in geotechnics to characterize the soil–water relation (Fredlund & Xing, 1994) allow the following of sample water contact in relation with the increase of applied capillary suctions. These curves are realized for various initial compacted samples and lead to the twinning identification of (i) the residual water content above which the path-connected space of the fluid phases is obtained within the compacted sample; it corresponds to the limit above which the water phase becomes funicular (liquid phase connection); and (ii) the inlet air water content. These trials consist of the establishment of equilibrium between capillary suction, water content, and apparent volume of the sample (Marshall et al., 1996).

3.6. Mapping the hydrotextural diagram

Plotting the specific water contents earlier defined in the hydrotextural diagram allows the identification of characteristic zones where the state of the two fluid phases is in relation with the compactness sample state (Fig. 2.8). Between the curves of liquid and gaseous percolation thresholds, the two fluid phases are simultaneously path-connected within the porous network. It is in this zone that the diphasic flows can occur. Taking into account the limit compactness of the solid phase (fluidization and jamming limits), all these different limits draw the closed surface corresponding to the simultaneous path-connection of the three phases (solid, liquid, and gaseous).

The detailed description of the tests, which identifies the granular medium state transitions of the wet microcrystalline cellulose, kaolin, and calcium phosphate, is developed in Ruiz, Rondet, Delalonde, and Desfours (2011) and in Rondet et al. (2009).

The hydrotextural diagram illustrates the different phase transitions essentially explained by the loss or gain in the constituents' path-connected space and by the different consistency states. The hydrotextural diagram describes the suitability of granular media to physicla changes. On this hydrotextural map, it will be necessary to add the influence of the shaping processes and their specific operating ability. The phase diagram is then

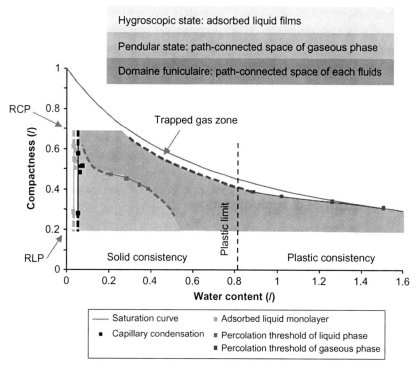

Figure 2.8 Microcrystalline cellulose hydrotextural diagram. (For color version of this figure, the reader is referred to the online version of this chapter.)

built by adding the "loading" axis, which refers to the characteristic thermo-dynamic quantity describing the applied process, that is, compression stress, mixing energy, relative humidity, etc.

4. AGGLOMERATION MECHANISMS AND AGGLOMERATE GROWTH MAPS

Mechanisms associated with the agglomeration of food powders are varied and complex (Saleh & Guigon, 2009) as wet agglomeration involves a large number of parameters of physical and physicochemical nature. The main goal is to increase the size of particles and generate agglomerates. Numerous bibliographical resources are available and summarize the state of the art (Ennis, Tardos, & Pfeffer, 1991; Iveson et al., 2001; Litster & Ennis, 2004; Palzer, 2011; Saleh & Guigon, 2009). During the wet agglom-eration processes, the addition of water to the dry native particles initiates the agglomeration mechanisms. The agglomeration process is conducted to manage two opposite mechanisms: attractive forces to elaborate the agglomerates versus rupture forces to disintegrate the agglomerates. The process parameters (shear rate, liquid flow rate, process time, etc.) control the contribution of the different agglomeration mechanisms. The description of wet agglomeration mechanisms is proposed based on six phases: wetting, nucleation, growth, consolidation, rupture, and stabilization (Iveson et al., 2001). Two phenomena are essential to the agglomeration of particles: (i) contact between individual particles; (ii) expression of their surface "reactivity" regarding stickiness and the formation of bridges (liquid or viscous) between particles coming into contact.

4.1. Fluidized bed agglomeration

Fluidized bed agglomeration consists of spraying a liquid onto or into a bed of initial particles (100–200 μm) fluidized with hot air, to render them locally sticky. Collision between wet "sticky" particles allows adhesion, with the formation of liquid and/or viscous bridges. Simultaneous drying by the hot fluidization air leads to the consolidation of the new structure formed. Agglomerates grow progressively due to the repetition of these different steps (mixing, wetting, collision, adhesion, and drying) until they reach sizes of 1–2 mm.

In fluidized bed agglomeration, contacts between individual particles arise due to their agitation, generated by the fluidizing air-flow. They are the result of either collisions between particles, if the bed density is low, or frictional contacts, if the bed density is high (low fluidized bed expansion). In any case, a fluidized bed is considered a low shear device (Schaafsma, Vonk, & Kossen, 2000).

The surface reactivity of particles is expressed by the combination of simultaneous thermal, hydrous, and mechanical constraints arising from the fluidization of particles by hot air and the wetting of their surface by the liquid sprayed. When particles are "inert," this liquid is a binder solution bringing the particle "reactivity" required for agglomeration with the deposit of binder drops on their surface. For most food powders, soluble in water, or containing compounds (carbohydrates) able to undergo glass transition in the conditions of fluidized bed agglomeration, the liquid sprayed is water. In both cases, the particle surface reactivity generated is due to its interaction with the liquid sprayed (wetting, sorption, dissolution, or plasticization). In the case of amorphous components, they become sticky when reaching the rubbery state, where the particle surface viscosity tends to a critical value of about 10^8 Pa s, corresponding to temperatures T_c

10–20 °C above T_g (Palzer, 2011). Due to the plasticizing effect of water, the glass transition temperature decreases when increasing the water content at the particle surface, allowing the sticky temperature T_c to be reached in the conditions of fluidized bed agglomeration (temperature, water content).

The degree of dispersion of the liquid sprayed in the powder bed is therefore one of the parameters controlling the agglomeration mechanism (Tardos, Khan, & Mort, 1997). It depends on the conditions of atomization, the liquid spray rate, and the particle bed fluidization. The liquid distribution in the spray zone can be characterized by the dimensionless spray flux Ψ_a, which is the ratio of the projected area of drops from the spray nozzle to the area of powder surface through the spray zone (Lister et al., 2002):

$$\psi_a = \frac{3\dot{V}}{2\dot{A}d_d} \qquad [2.4]$$

where $\dot{V}$ is the volumetric spray rate (m^3 s^{-1}), d_d is the drop diameter (m), and $\dot{A}$ is the flux of powder surface traversing the spray zone (m^2 s^{-1}). ψ_a values calculated in the early stages of fluidized bed agglomeration are very low, indicating a good dispersion of the sprayed liquid within the particle bed.

Although the fluidized bed provides intense mixing, the air temperature and humidity recorded throughout the particle bed during agglomeration lead to the consideration of three regions in the well-mixed fluidized bed (Fig. 2.9), with sizes depending on the operating conditions (Heinrich, Blumschein, Henneberg, Ihlow, & Mörl, 2003; Jimenez et al., 2006):

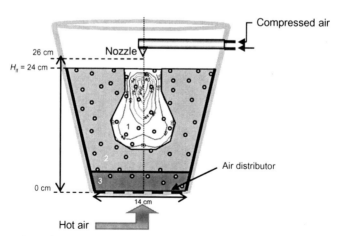

Figure 2.9 Three thermal zones in a conical top-sprayed fluidized bed. *Adapted from Jimenez, Turchiuli, and Dumoulin (2006). (For color version of this figure, the reader is referred to the online version of this chapter.)*

i. The wetting-active zone, below the spraying nozzle at the topmost part
of the bed. It is a low temperature and high humidity region, characterized by high humidity and temperature gradients due to the wetting
of the fluidized particles by the sprayed liquid and the evaporation of
the solvent.

ii. The isothermal zone, near the walls and around the wetting-active
zone, where there is equilibrium between heat and mass transfer and
air temperature is homogeneous.

iii. The heat transfer zone, situated right above the bottom air distributor
plate. In this narrow area, the hot air temperature decreases strongly due
to the energy absorbed by the colder particles coming from the upper
zones.

The spray zone where liquid drops collide with moving particles is thought
to correspond to the wetting-active zone, the shape, volume, and depth of
which depend on the operating conditions. Top spraying water in a fluidized
bed of glass beads, Jimenez et al. (2006) estimated that in conditions leading
to controlled agglomeration, this zone occupied 22–29% of the fluidized bed
volume with a penetration depth of 12–18 cm (Fig. 2.9). Assuming a homogeneous distribution of particles within the whole fluidized bed, this fraction
of the bed volume corresponds to the fraction α of the total number of particles in the spray zone at time t.

In fluidized beds, the particle motion is complex, and it is often assumed
that particles are dragged in the trail of the rising air bubbles. The upward
mass flow rate Q_p of particles in the fluidized bed (kg s^{-1}) can then be
obtained from the average particle circulation time τ_c (s) and the bed particle
load M_p (kg):

$$Q_p = \frac{M_p}{\tau_c} \qquad [2.5]$$

Agglomerate growth occurs in two stages, with the first, a rapid increase
of the median diameter, starting either as soon as spraying starts or after a
delay for "inert" particles (sufficient binder deposit), and the second, a slow-
down till reaching a plateau corresponding to equilibrium between growth
and breakage due to collisions/friction (Fig. 2.10).

The growth rate in the first stage depends on the operating parameters:
liquid flow rate, particle load, drying air temperature and flow rate, sprayed
liquid properties, and drop size (Fig. 2.11A). But, for milk powder agglomeration, it was found that agglomerate growth mainly depends on the
amount of water sprayed whatever the flow rate, the particle load, and even

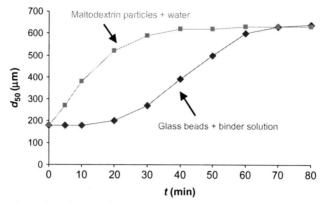

Figure 2.10 Typical evolution of agglomerate diameters during fluidized bed agglomeration. (For color version of this figure, the reader is referred to the online version of this chapter.)

the nature of the milk particles used (skim or whole) in the conditions studied (Fig. 2.11B) (Barkouti, Turchiuli, Carcel, & Dumoulin, 2012). The evolution of the particle size distribution showed that, during the first stage, initial particles associate, giving rise to an intermediate population (250 µm) which disappears progressively during the second stage to form agglomerates growing slowly from an initial size of about 400 µm to a final mean diameter of about 700 µm (Fig. 2.12). This growth mechanism leads to three typical evolutions of the mass fraction in each size class with time (Fig. 2.12B and C): (1) fast decrease during the first minutes for initial particles; (2) fast increase during the first minutes and then progressive decrease for the intermediate population; and (3) progressive increase for agglomerates.

This is consistent with a drop-controlled regime corresponding to low Ψ_a values where agglomeration occurs in stages, with first, the association of initial particles into small initial structures, the size of which depends on drop size, from which larger agglomerates are progressively built (Fig. 2.13) (Hapgood, Litster, White, Mort, & Jones, 2004).

4.2. Low shear mixer agglomeration

The results presented in this section are a synthesis of the heuristic model derived from the analysis of experiments in low shear mixer on several powders wetted by deionized water: kaolin, microcrystalline cellulose, dicalcium phosphate, and semolina.

For a given wetting–mixing stage (residence time, load, rotational speed, water content, and binder spraying conditions), the agglomerate populations

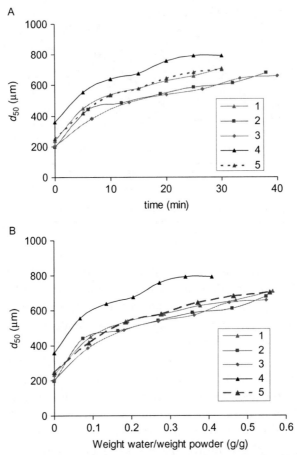

Figure 2.11 Evolution of median diameter during agglomeration for skim (curves 1–4) and whole (curve 5) milk powder as a function of time (A) and quantity of water sprayed (B). (For color version of this figure, the reader is referred to the online version of this chapter.)

obtained are considered according to their dimensional and hydrotextural characteristics (solid volume fraction, water content and saturation degree at bed, and agglomerate scales).

The size distribution curves are obtained by sieving (Fig. 2.14A) and provide the evolution of the median diameter according to the water content (Fig. 2.14B) to which it is possible to associate a distribution criterion. For all the products studied, the solid volume fraction assessed at the agglomerate scale lies below the saturation curve (Fig. 2.15A) indicating that agglomerates are unsaturated structures. This trend is confirmed for different blade

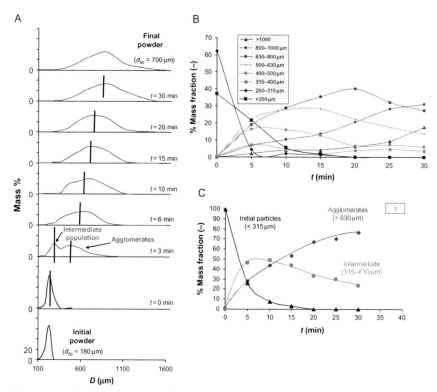

Figure 2.12 Typical evolution of the particle size distribution during milk powder fluidized bed agglomeration. (For color version of this figure, the reader is referred to the online version of this chapter.)

Initial   Intermediate   Agglomerates

Figure 2.13 Agglomerate growth mechanism in fluidized bed. (For color version of this figure, the reader is referred to the online version of this chapter.)

rotational speeds, the shear remaining at a "low" level (Fig. 2.15B). It is possible to observe that agglomerate solid volume fraction decreases with water content until the saturation curve is reached for a water content very close to the plastic limit (w_p) defined in the sense of Atterberg limit (see Chapter 3). During their progressive saturation, agglomerates expand, indicating that they do not consolidate when the water content increases.

From each water level, it is possible to establish a population balance of each "structure" in the bed bulk (Fig. 2.16). The identification of native

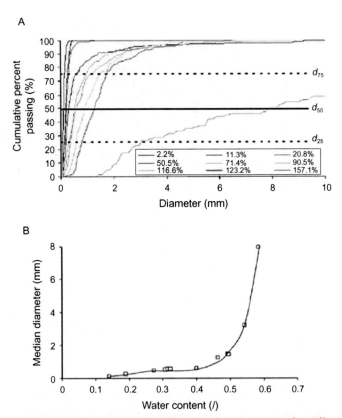

Figure 2.14 (A) Evolution of the size distribution of agglomerates for different water contents. Case of microcrystalline cellulose. (B) Evolution of the median diameter with increasing water content. Case of semolina. (For color version of this figure, the reader is referred to the online version of this chapter.)

semolina, nuclei, agglomerates, and dough pieces is based on the twin analysis of structure size and hydrotextural parameters. For example, two structures of apparently the same diameter can be distinguished by their solid volume fraction: the structure with the higher solid volume fraction is a nucleus if w is low or an agglomerate if w is high. The same methodology is used to distinguish between agglomerates and dough structures. The population balance makes it possible to delineate an agglomeration water range.

4.2.1 Wetting/nucleation
Nucleation occurs in the wetting area of the mixer bowl. The particle bed is sprayed with droplets that moisten the grain and create capillary bridges between particles. The first structures formed have a rather spheroidal shape

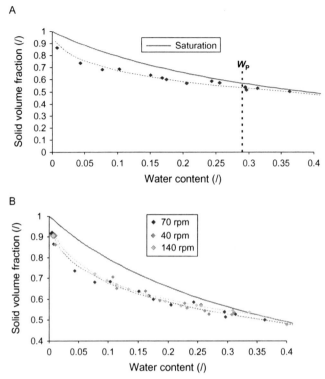

Figure 2.15 Hydrotextural diagram for agglomerates. (A) Evolution of the agglomerate size distribution with water contents. Case of kaolin. (B) Influence of the blade rotational speed. (For color version of this figure, the reader is referred to the online version of this chapter.)

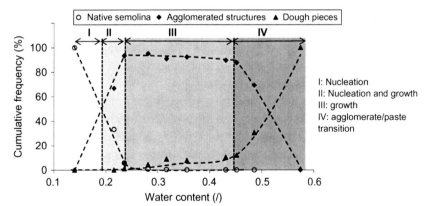

Figure 2.16 Example of population balance of the different structures coexisting in a wetted powder bed for different water contents. Case of semolina. (For color version of this figure, the reader is referred to the online version of this chapter.)

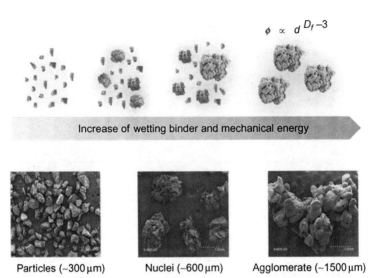

$$\phi \propto d^{D_f - 3}$$

Increase of wetting binder and mechanical energy

Particles (~300 μm)   Nuclei (~600 μm)   Agglomerate (~1500 μm)

Figure 2.17 Schematization of the agglomeration of durum wheat semolina in couscous. (For color version of this figure, the reader is referred to the online version of this chapter.)

and small size (of the order of magnitude of two grain diameters) and are very compact (Fig. 2.15). These compact and small structures are nuclei (Fig. 2.17). They are the "seeds" of agglomerates, which are larger and less compact (Figs. 2.15 and 2.16). Many factors have an influence on this wetting/nucleation stage. The wettability of particles, first of all, is essential to promote capillary interaction. A "good" wettability allows liquid to access all sites of the particle surface likely to contribute to the establishment of capillary bridges. These sites are located close to the intergranular contacts. When the wettability is partial, capillary bridge-building is much more dependent on the ability of the process to redistribute the liquid within the powder bed. However, "good" mixing of a granular medium is a delicate operation that is not so obvious to carry out.

Wetting liquid viscosity is another important parameter that is involved in the absorption of the droplet fall on the powder bed, and in the inerting of drops between grains that promote the stability of capillary forces. The less viscous the fluid, the more it will be subject to a quick migration between grains. The surface tension of the liquid also plays an important role. Nuclei are more porous when surface tension is high and vice versa. Indeed, when surface tension is high, the intensity of the capillary force is directly proportional. This capillary force is thus intense enough to allow the maintenance of less compact arrangements between particles, which may resist the

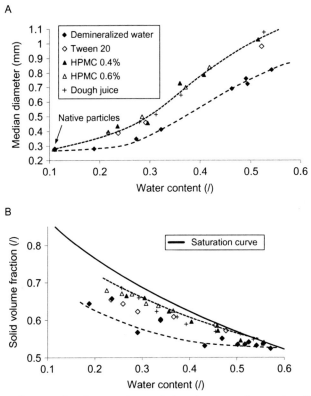

Figure 2.18 Influence of surface tension and viscosity on agglomeration. Case of semolina. (A) Agglomerates size. (B) Hydrotextural diagram.

rupture forces induced by the mixer blade. When surface tension is low, nuclei are much more compact, allowing them to resist the frictional shear despite weaker friction forces between particles (Fig. 2.18B). Through this example, it is possible to see that the effects of the interaction of these two factors are difficult to predict and significantly counter-intuitive.

Indeed, for semolina, the dissolution of proteins and polysaccharides in the liquid-wetting induced kinetic effects, which reduced surface tension and increased viscosity. Variations in surface tension and viscosity of the binder are obtained by the formulation of deionized water with conventional agents (Tween and HPMC) (Barkouti, Rondet, Delalonde, & Ruiz, 2012). The ranges of surface tension and viscosity tested are depicted in Table 2.2. "Semolina juice" helps to overcome kinetic aspects related to the reactivity of surface particles. Figure 2.15 depicts the area of curve sizes likely to be obtained for semolina agglomerates in low shear mixer.

Table 2.2 Physicochemical properties of binders

Binders	γ (mN m^{-1})	μ (mPa s)	V^* (m s^{-1})
Water	71 ($\pm$0.1)	1.0 ($\pm$0.02)	71.0 ($\pm$1.5)
Semolina "juice"	35 ($\pm$0.1)	1.5 ($\pm$0.10)	23.3 ($\pm$1.6)
Tween 20	35 ($\pm$0.6)	1.0 ($\pm$0.03)	35.0 ($\pm$1.6)
HPMC (0.4%)	46 ($\pm$2.4)	14.8 ($\pm$0.20)	3.1 ($\pm$0.2)
HPMC (0.6%)	46 ($\pm$1.1)	44 ($\pm$2.00)	1.0 ($\pm$0.07)

γ, surface tension (in mN m^{-1}); μ, dynamic viscosity (in mPa s); V^*, characteristic speed ($V^* = \gamma/\mu$) (in ms^{-1}).

The reactivity of the powder surface is essential for the modulation of the agglomeration process. Hydration can lead to gelatinization or cementing, which adds cohesion between particles. If the characteristic time involved in the kinetics of surface reaction is of the same order of magnitude as that of the duration of the operation, these phenomena can become predominant in agglomeration mechanisms. On the other hand, when the duration of the operation is short, the capillary interaction is the major structuring force that ensures bonding of the particles.

In this situation, one of the key parameters is the relationship between the size of the drops and that of grains. Depending on the value of this ratio, two characteristic regimes of nucleation can be distinguished. When the drops are smaller than the grains, they are coated by liquid film. They become sticky and associate before they contact the blade that tests the cohesion of the structure they form. This association generates agglomerates interacting thanks to the action of capillary menisci renewed as they pass under the spray. When the drops are larger than the grains, they are immersed and constitute nuclei whose size is directly dependent on the drop diameter. These nuclei, looser and more cohesive, can be broken by the blade and generate agglomerates able to collect dry particles on their surface and interact with other nuclei by plastic interpenetration. When they are generated by dispersion of fine droplets, nuclei have a high solid volume fraction, close to the RCP, and a low liquid content. On the other hand, the immersion regime leads to less compact and more humid nuclei.

Nucleation is the first prestructuration step during which the particles interact with the drops. The powder surface reactivity, along with the properties of the binder drops as well as the wetting, homogenization, and frictional conditions, makes it possible, according to their respective modulation, to generate several kinds of nuclei which induce the formation of agglomerates whose size, shape, and texture can be very different.

4.2.2 Growth

The mechanical mixing conditions will play an essential role in the association of nuclei and agglomerates that lead to the structures' growth. The role of the mixer is multiple.

First of all, it allows a homogenization of the granular medium in the bowl. The effectiveness of the mixture is crucial because several granular populations coexist: the native particle median diameter (d_{50}^*), nuclei ($2–3 \times d_{50}^*$), agglomerates ($5–10 \times d_{50}^*$), and dough pieces (beyond $10 \times d_{50}^*$). These orders of magnitude indicate that the agglomeration is a method that permits the increase of the size by approximately a decade. As a comparison, aggregation of proteins, polymers, and colloids leads to structures whose size is greater than 10–100 that of native particles. For granular media, this limitation of size extension is partly due to the fact that, when the size of the bonded objects is of the order of magnitude of the capillary length, gravity outweighs the capillarity, and on the other hand, the liquid quantity necessary to the achievement of these sizes is such that the agglomerate rheological behavior becomes plastic and thus favors their association in pasty structures. The structural "integrity" of the agglomerates is no longer ensured.

Then, when one tries to mechanically homogenize a polydispersed particle population, the phenomenon of segregation by size occurs. Larger particles are found on the surface of a bed constituted of small size particles. The wetting area being located at the surface of the bowl, the growth of the larger particles would be more pronounced if the mixer does not limit segregation.

The mixer must also ensure sufficient renewal of particles at the bed surface to contribute to an efficient dispersion of the sprayed drops. The residence time of particles, nuclei, and agglomerates at the surface should be long enough to allow their hydration and short enough so that water does not migrate too deep into the bed.

Finally, the shear stresses generated by the blade must be transmitted to the agglomerates to test their mechanical strength. The agglomerates that directly pass in the vicinity of the blade are sheared off with an energy that depends on the rotational speed and its distance range. The other agglomerates, stirred beyond the direct contact of the tool, rub between them, and it is the inter-agglomerate friction that allows testing the mechanical strength of the structures. The network of forces generated in the vicinity of the blade makes it possible to (i) consolidate agglomerates that are located at the edges of the bowl, (ii) destroy agglomerates whose mechanical strength is lower than the shear strength, and (iii) shape cohesive agglomerates in the form of spheres.

Therefore, the mixer plays a role equivalent to that of a "barrier of potential." Agglomerates whose cohesion is high can pass this barrier; the others turn back to the nuclei state so as to begin the growth stage again. Despite the apparent roughness of agglomerates that can be of the order of magnitude of the grain size, their geometry can be compared to a sphere whose diameter is equivalent to the largest size. During the continuous supply of the binder, the growth of the agglomerates, which results in an increase in the diameter median population, (Figs. 2.14 and 2.18) is concomitant with the filling of air voids (Fig. 2.19). The increase in the degree of saturation according to water content can be modeled by a power law (Rondet, Delalonde, Ruiz, & Desfours, 2010):

$$S = \left(\frac{w}{w_{sat}}\right)^n \quad \text{with } n < 1 \text{ and for } 0 < w \leq w_{sat} \qquad [2.6]$$

where w_{sat} is the saturation water content and n is a parameter related to the deformability of the medium. This parameter reflects the hydromechanical coupling, that is, the influence of the presence of water in the granular medium on the hydrotextural state. When the medium is not deformable, it is shown that $n = 1$, otherwise $0 \leq n < 1$. Eq. (2.6) generalizes the definition of the degree of saturation.

In an unsaturated wet granular medium, water content, compactness, and saturation degree are linked by a conservation balance (Rondet, Delalonde, et al., 2010):

$$\phi^{-1} = 1 + d_s^* \frac{w}{S} \qquad [2.7]$$

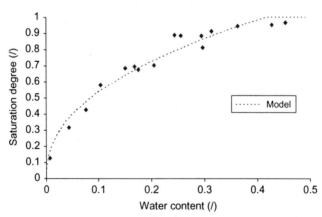

Figure 2.19 Variation in the saturation degree of agglomerates with water content. Model (cf. Eq. 2.6) vs. experiments (case of kaolin). (For color version of this figure, the reader is referred to the online version of this chapter.)

By considering Eq. (2.6) and Eq. (2.7), the relationship between the compactness, the water content, and the degree of saturation becomes:

$$\phi^{-1} = 1 + d_s^* w_{sat}^n w^{1-n} \qquad [2.8]$$

With this relationship, the mean hydrotextural state of agglomerates can be set from a single state variable: the water content.

The association of dense particles leads to larger structures that incorporate interparticle space as "void" in their volume. When measuring simultaneously the changes in the size as well as the solid volume fraction of the structures (Fig. 2.20) at the same time, it can be shown that these two parameters do not vary independently. Indeed, during the growth stage, the median diameter is always correlated to the compactness. The agglomerates grow, resulting in an increase of the d_{50}, but at the same time they become less and less compact. For each water content, the median diameter and the solid volume fraction are related by a power law (Rondet, Delalonde, et al., 2010):

$$\phi = \phi_N \left(\frac{d_{50}}{d_{N50}} \right)^{D_f - 3} \qquad [2.9]$$

where D_f is the fractal dimension of the agglomerate, 3 is the dimension of the space in which the object is immersed, ϕ_N is the solid volume fraction of the nuclei, and d_{N50} their median size. The fractal dimension is between 2 and 3. The more it tends toward 3, the more compact is the structure. Generally, $2.5 \leq D_f \leq 2.8$ for all agglomerates processed in low shear mixer (Rondet, Ruiz, Delalonde, Dupuy, & Desfours, 2010). The fractal dimension

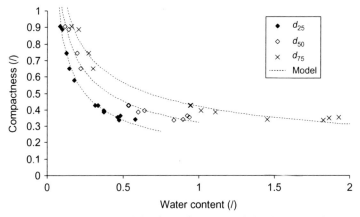

Figure 2.20 Relation between solid volume fraction and the diameter of agglomerate populations (case of microcrystalline cellulose). Model using Eq. (2.7).

of the agglomerate represents its own dimension. The agglomerate is a porous structure obtained from a growth mode that provides a similarity whatever the observation scale. Zooming in on the structure makes it possible to find the same topological nature at every scale between the nuclei (basic pattern) and the maximum size of the object. These two sizes (nuclei size and the maximum size of the object) represent the two correlation lengths between which is set a physical fractal (Jullien & Botet, 1987; Sander, 1986). In the case of granular media, this size ratio does not exceed a decade. This is a limit case of fractal structure (2–3 decades occur commonly in microscopic objects consisting of macromolecules and nanoparticles, or in metric objects such as clouds, for example).

This heuristic model, which reflects the agglomerate texture, requires the identification of five independent parameters: $\{n, w_{sat}, D_f, d_{N50}, \phi_N\}$. It should be noted that the power law between the diameter and the solid volume fraction has been validated for all other quartiles, indicating that all the agglomerates of the distribution grow in the same way, the fractal dimension being the same for all the quartile (Fig. 2.20).

Thus, agglomerates grow by the association of nuclei that incorporate "voids" between them. This type of association is not unique. Indeed, this feature (size increase concomitant with a decrease of solid volume fraction) is common to many physical or biological systems such as colloidal aggregates, polymer aggregates, snowflakes, trees, etc. (Sander, 1986). If the diameter is characteristic of the "external" form of the agglomerates, compactness represents the topological nature of the "internal" form of the object. The inside and the outside are therefore correlated during the elaboration process. The physical processes that lead to the elaboration of an agglomerate by the successive association of substructures under the twin action of forces, statistically distributed, and a local attractive interaction, induced a fractal structure. These processes are induced by stresses that are external to the granular system: the sprayed liquid binder and the mechanical energy.

4.2.3 Agglomerates/paste transition

The population balance (Fig. 2.16) indicates that the agglomeration regime gradually ends with the association of agglomerates in increasingly larger and plastic structures. Then, with the accurate water content, they "merge" into a continuous dough. It can be shown that the median diameter growth of pasty structures follows a power law according to the water content (Fig. 2.21). This growth pattern is characteristic of the percolation phenomenon (Guyon, 2005; Stauffer & Aharony, 1994):

$$d_{50} \propto (w - w_T)^p \qquad [2.10]$$

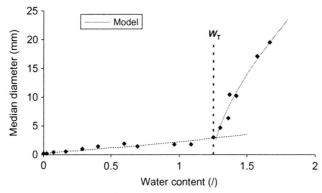

Figure 2.21 Agglomerate/paste transition. Comparison between model (using Eq. 2.8) and experiments (case of microcrystalline cellulose). (For color version of this figure, the reader is referred to the online version of this chapter.)

where w_T is the transition water content and p is a critical exponent. Once the homogeneous dough is saturated with liquid, growth follows a size increase that only depends on the mass of dry particles originally introduced (m_s) and water content. This can be assessed by calculation:

$$d_{50} = \left[\frac{6m_s}{\pi \rho_s^*} \left(1 + d_s^* w\right) \right]^{1/3} \qquad [2.11]$$

The values of the power law exponent (Eq. 2.8) are relatively close for the same product, regardless of the operating conditions. It is possible to conclude that the agglomerate/paste is a percolation phenomenon whose threshold is w_T. The low variability of the parameters of Eq. (2.8) makes it possible to say that this percolation is independent of the operating parameters and appears to be only sensitive to the nature of the product.

5. AGGLOMERATION PROCESSES

For food applications, different agglomeration technologies have to be applied to combine fine primary particles to larger ones (Fig. 2.22). The technologies can be classified into two categories: pressure agglomeration and wet-controlled growth agglomeration (Litster & Ennis, 2004; Palzer, 2011; Saleh & Guigon, 2009).

5.1. Agglomeration under static pressure

Agglomeration under static pressure refers to processes in which the powder is transformed into compact form by simple compression. Static pressure is applied to confined powder to generate adhesion between the particles. The

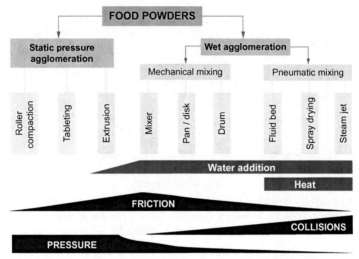

Figure 2.22 Diversity of agglomeration technologies of food powders.

almost spherical native particles have to be deformed under pressure to decrease the distance between their surfaces and generate contact area. When pressure agglomeration is conducted without liquid addition, van der Waals forces are responsible for particle adhesion. Dry agglomeration is of particular interest, as the final product requires no drying process. A liquid binder can be incorporated to increase the adhesion forces between the particles. Depending on the nature of the material, the compression stage can be followed by an additional sintering stage to stabilize the contact points. It can be operated using different technologies: roller compaction, tableting, or extrusion (Fig. 2.22) (Palzer, 2011).

5.2. Wet agglomeration

Wet-controlled growth agglomeration refers to agglomeration processes during which a liquid binder is pulverized over an agitated powder bed. This type of agglomeration is also called "collision agglomeration" or simply "wet agglomeration." It is a complex process, which involves a large number of parameters of physical and physicochemical natures. Whatever the powder type, the wet agglomeration process is based on the coupling of two unit operations: (1) liquid addition to develop adhesion forces between the particles and (2) mixing of the powder bulk to disperse the liquid over the particles and promote the growth mechanisms by enhancing different typologies of motions (e.g., shearing, compression, rotation, translation, etc.) and by controlling kinetic energy of the colliding particles and the contact time of collision.

The agglomeration equipment is able to both agitate the particles and introduce the liquid binder on the moving particles to initiate the agglomeration mechanisms. A final drying stage is required to stabilize the agglomerates. The agglomeration equipment is classified into two categories according to the modalities of particle agitation (Fig. 2.22) (Palzer, 2011; Saleh & Guigon, 2009).

The agglomerators with pneumatic mixing use an air stream to agitate the particles under low shear conditions and to dry the agglomerates. The drying stage is then conducted almost simultaneously with the agglomeration stage. The pneumatic mixing processes are complex because many competitive stages occur simultaneously: powder mixing, liquid spraying, particle wetting, agglomeration, and drying. The liquid is directly sprayed on the agitated particles. The necessary heat to dry the agglomerates is brought by the heated air stream. Three different agglomerators with pneumatic mixing are used for food applications: steam jet, spray drying, and fluid bed.

The agglomerators with mechanical mixing use mixers to agitate the particles. They can be classified into two categories (with rotating walls or rotating blades) (Saleh & Guigon, 2009). The mixers with rotating walls (drum, pan, or disk agglomerators) only generate moderate shear rates. The mixers with rotating blades (called mixer agglomerators) can generate high or low shear rates, depending on the geometrical characteristics of the mixers and the blade rotation speeds. Different mixer agglomerators exist, according to the position of the blade (vertical, horizontal, or oblique) and geometries of the tank and blade. The liquid is directly sprayed over the powder bulk under mixing conditions. Mixer agglomerators can work with powders of large size distribution and allow a good distribution of very viscous liquids (Knight, Instone, Pearson, & Hounslow, 1998). The final drying stage is then conducted, after the agglomeration stage.

5.3. Powder flows and stress transmission in the powder bed of mixer agglomerators

During wet-controlled agglomeration using mechanical mixing, powder flows directly and participates in the growth and rupture mechanisms through the generation of particle–particle and particle–mixer contacts. Despite the lack of constitutive laws (equivalent to the fluid rheology) that would allow the description of powder flows, three dry flowing regimes have been identified: quasi-static, frictional, and collisional (Pouliquen & Chevoir, 2002; Rajchenbach, 2003). Depending on the stress patterns induced by the agglomerator, these flowing regimes are present at specific intensity levels and are possibly differently localized in the powder bed. It can be noted that the nature of the flow is likely to vary with water addition due to modifications of the powder properties (at the particle, agglomerate, and bed scales).

In the case of mixer agglomerators, it is necessary to better understand the stress transmission modalities from the blade to the powder, which are at the origin of powder flows. The capacity of a mixer to agglomerate hydrated particles depends on its aptitude (i) to induce a stress field in the powder bed and (ii) to uniformly transmit the stresses at the granular medium, in order to enhance the contact and adhesion between hydrated particles through capillary bridges. The control of the wet agglomeration process in a mixer agglomerator requires access to a better description of the powder behavior during its flow, and the effect of binder addition to the flow patterns. In this process, particles are directly in contact, and the powder behavior is governed by friction and interparticular forces, cohesion (partly linked to capillary forces in the case of a wet granular medium), gravity, and forces induced by motions against side walls (particle–mixer wall friction) (Condotta, 2005).

The growth models of the literature allow us to predict the maximum size accessible to agglomerates at the end of the process and the effect of different variables on this growth. However, none of these models is industrially used to control the wet agglomeration process (Iveson et al., 2001). This is linked to the difficulty in defining the model parameters, more particularly those related to powder rheology, frequency, and velocity of collisions. The discontinuous nature of powders and the existence of several scales over which the reaction takes place prevent the direct use of the constitutive laws of continuous media and thus the determination of theoretical velocity profiles in the mixer. Describing and predicting the stress and velocity distributions in a granular medium constitute a challenging issue. The analysis of velocity and trajectory fields is essential to accurately characterize granular flows. The acquisition and treatment of these fields allow us to identify typologies of particle motion at three scales (at the particle or "micro" scale, at the intermediary scale of a small assembly of particles or "meso" scale, and at the whole powder bed or "macro" scale), and then to define pertinent physical quantities to characterize the flow (e.g., length of a moving zone).

Generally, approximated by the blade tip velocity, the particle impact velocity can be determined experimentally. Some techniques such as positron emission particle tracking and high speed images acquisition (particle image velocimetry (PIV) technique) seem very promising to get access to the powder behavior (velocity gradients and particle motions). The PIV technique has been employed frequently in the last 10 years for the noninvasive study of powder flows in different configurations. PIV was used to characterize surface velocities in low and high shear mixers (Lister et al., 2002; Muguruma, Tanaka, & Tsuji, 2000; Nilpawar, Reynolds, Salman, & Hounslow, 2006; Russell, Diehl, Grinstead, & Zega, 2003; Wellm, 1997).

Other studies related to the quantification of granular stresses and displacement fields have been conducted by physicists studying granular matter. They consist in measuring the drag forces exerted on an intruder moving slowly in a granular bed (Albert, Pfeifer, Barabasi, & Schiffer, 1999; Geng & Behringer, 2005), quantifying the influence of an intruder motion on velocity fields in a monolayer of grains (Candelier & Dauchot, 2009), or evaluating the influence of a flat blade moving horizontally in a homogeneous granular bed constituted of polydisperse glass beads (Gravish, Umbanhowar, & Goldman, 2010). By giving new insights into the mechanical properties of granular media subjected to a moving blade, these fundamental scientific studies will surely contribute to a better understanding of phenomena contributing to powder agglomeration mechanisms. The study of stress transmission in a powder bed is an emerging field of interest, which will help to improve the modeling aptitude of agglomeration and provide data for the development of well-adapted equipment and process control. However, to allow a transposition of results obtained on pilot mixers with model powders to industrial mixers with natural powders, work must also integrate the food powder complexity by applying a coupled physics--technology approach. The work of Mandato, Cuq, and Ruiz (2012) was mainly conducted on complex food powders (durum wheat semolina and wet agglomerates) and on a model powder (glass beads) traditionally used by physicists and mechanics specialists. Characterization elements of a powder bed were obtained by the study of mechanical input contributions to dry powder flow. It represents the first step, before the wet and cohesive powder flow description. A simplified model mixer associated with a unidirectional blade was developed to analyze the system on two dimensions (Fig. 2.23A). The initial static state of a laterally confined powder bed was first described, to extract pertinent criteria to help understand dynamic phenomena observed under low shear mechanical stresses. The work of Mandato (2012) is included in a relatively little studied field as most of the agglomeration studies deal with the agglomeration process in a collisional regime under high shear conditions. The "collision" concept is then entirely appropriate. In a frictional regime, the flow is localized in shear banding areas and particles are almost always in interaction such that labile force networks are constantly created and broken. The visualization of the

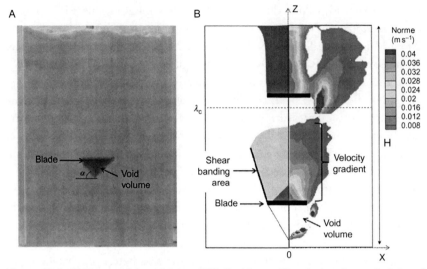

Figure 2.23 (A) Picture of the mixing cell filled with semolina during the vertical rise of the blade. (B) Representation of instantaneous velocity fields for two different heights of the blade during its vertical rise in semolina at $v = 40$ mm s^{-1} (right) and scheme of the different zones and patterns observed in PIV (left). (For color version of this figure, the reader is referred to the online version of this chapter.)

behavior of nonhydrated grains in the neighborhood of a moving blade allows the identification of different types of particle motions in two dimensions (compaction, dilatation, loading and rupture of the chain forces, lateral collapsing, avalanches, shear bands, etc.) depending on the distance to the blade, the blade velocity and depth, and the powder properties (e.g., compactness, size polydispersity, etc.) (Gravish et al., 2010; Mandato, 2012). The intruder motion induces a velocity gradient whose vertical and horizontal extents characterize the stress transmission outreach (as illustrated in Fig. 2.23B). Concerning the results of Mandato (2012), a frictional regime is observed below a characteristic depth (already identified under static conditions) and a "low-frustrated" flowing regime in the zone between the characteristic depth and the surface. The dissipation of the mechanical energy provided by the blade is thus dependent on the blade penetration depth in the powder bed. When the blade moves in a depth inferior to the characteristic depth, a part of the mechanical energy is transmitted to the grains and allows powder flow, which can lead to an increase in the collision probability of particles and drops, enhancing the agglomeration process. The other part of the energy is dissipated by intergranular friction. This proportion of dissipated energy significantly increases when the blade depth is higher than the characteristic depth. Moreover, in this frictional zone, the probability of ruptures of agglomerates is higher due to abrasion. In the context of wet agglomeration, the characteristic length can be useful to design bowl geometries and to assure a bowl height lower than its width. For a mixing bowl of defined dimensions, the characteristic length can also be used as a reference regarding the filling rate.

Thus it appears necessary to consider distributions (of particle velocities and shear) and not only the mean values to describe the behavior of granular media under mechanical solicitation. A study at the intermediate scale ("meso") allows obtaining distributions of shear and velocities at the powder bed scale. It is then possible to define pertinent criteria related to the agglomeration mechanisms, namely, a volume of powder really put in motion by the blade, and above all to obtain the description of typologies of particle motions. The characteristic extent of a velocity gradient needs to be quantified. The blade speed modifies the intensity ranges of particle velocity and the characteristic lengths of motion fields. Technologies and associated control parameters modulate the typologies of particles' motions and by cause-and-effect link are supposed to affect the agglomeration mechanisms. The blade is not any longer considered a particle collision promoter. Through the created shear bands, the blade imposes a "potential barrier" that only the agglomerates stable enough and mechanically resistant can overcome. The dynamic study of Mandato (2012) shows that the energy consumed by the blade is lower in the upper zone (i.e., between the characteristic depth and the surface) of a powder bed than in the lower zone, where the level of friction is higher. It should thus be possible to entirely rethink the agglomeration processes by testing, for instance, other mixer geometries (e.g., through the ratio between the mixer height and diameter).

Additional work related to stress transmission needs to be done on hydrated food powders. The particle swelling properties may modify the stress transmission ability in the granular medium (through a modification of cohesion and plasticity) and thus modulate the volume of the blade.

5.4. Agglomeration sensors

The management of the agglomeration process is still mainly based on empirical knowledge and the technical know-how of operators. This process has been traditionally monitored by visual and sampling procedures. Physical and hydrotextural changes (e.g., particle size, water content, and compactness) of agglomerates are measured by different analytical methods according to the mixing time and/or the hydration level. Other methods are used to follow the agglomeration process in-line by using *in situ* sensors or probes, without off-line measurements (e.g., power and torque measurements, near infrared (NIR) spectroscopy). The use of sensors to follow the agglomeration process in-line has several goals: to characterize the evolution of the product during the process, contribute to the understanding of mechanisms, control the process efficiency, and suggest criteria for scale-up (Bardin, Knight, & Seville, 2004; Faure, York, & Rowe, 2001; Knight, 2001; Leuenberger, 2001; Malamataris & Kiortsis, 1997). The discontinuous nature of the granular medium makes the tracking of agglomeration particularly difficult to conduct.

The measurement of blade torque or motor power consumption constituted the first in-line method of agglomeration process control (Lindberg, Leander, Wenngren, Helgesen, & Reenstierna, 1974; Travers, Rogerson, & Jones, 1975). These methods are still the most developed today. Energy consumption measurements are often carried out for increasing water content and the obtained typical curve is considered in relation to variations in the different water states in the powder or to powder bed compactness variations. Recently, an evaluation of energy consumption in the wet agglomeration process was conducted for different hydration levels of durum wheat semolina (Rondet et al., 2012). Under the assumption that the mechanical and magnetic dissipation are constant, it was highlighted that it is possible to approximate, except for the heat dissipation, the energy consumed at the different levels (equipment, load, and agglomeration mechanisms). The total energy which is consumed by the blade rotation motion (not dissociated from the magnetic and mechanical dissipation) is considered the sum of energy consumed by the mixing and shearing effects of the mass load, by the agglomeration mechanisms, and by the cohesive behavior of the agglomerated products. The energy necessary for the agglomeration of semolina is relatively small, compared with total energy consumption (Fig. 2.24). It requires, at the most, only a third of the provided energy. The remaining energy is used for blade rotation and mixing of the load mass, or is lost through mechanical, magnetic, or heat dissipation. The energy measurements carried out by Rondet et al. (2012) give orders of magnitude to improve discussions about an ecodesign approach. This methodology could be considered an original method for the establishment of energy audit for the wet agglomeration process.

Among the other existing techniques, NIR spectroscopy appears to be a useful noninvasive tool to extract physical and chemical information about powdered products during an ongoing manufacturing process, particularly in the pharmaceutical and food industries (Alcalà, Blanco, Bautista, & González,

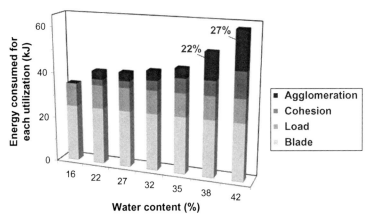

Figure 2.24 Impact of water addition level on energy repartition during agglomeration of durum wheat semolina in the presence of water. *Adapted from Rondet et al. (2012)*. (For color version of this figure, the reader is referred to the online version of this chapter.)

2009; Bertrand & Dufour, 2000; Dowell & Maghirang, 2002; Frake, Greenhalgh, Grierson, Hempenstall, & Rudd, 1997; Kaddour & Cuq, 2009; Miller, 1979; Osborne, 1999, 2000, 2006; Rantanen & Yliruusi, 1998; Robert, Devaux, Mouhous, & Dufour, 1999; Wellner et al., 2005). Only a few studies have been carried out to evaluate the ability of NIR spectroscopy to monitor powder agglomeration in-line (Alcalà et al., 2009; Frake et al., 1997; Kaddour & Cuq, 2009; Rantanen & Yliruusi, 1998). Frake et al. (1997) demonstrated that NIR spectroscopy could be useful to follow in-line the increase in water content and size of particles during a granulation stage in a fluidized bed. The work of Kaddour and Cuq (2009) and Mandato (2012) has already demonstrated the high potential of NIR spectroscopy to follow in-line the wet agglomeration process of cereal-based products, and more particularly, wheat-based food powders, with the possibility of describing the agglomeration or breakage mechanisms. These studies also reveal the usefulness of NIR spectroscopy in describing the physical and physicochemical changes occurring during the wetting and mixing stages of the agglomeration process. More particularly, analyses of spectral variations can be useful to discriminate two different processes under two different water supply conditions (i.e., rapid vs. slow supply), which are associated with two different kinetic evolutions of agglomeration mechanisms inducing specific changes in size distribution and water content per size fraction (Mandato, 2012). Different characteristic times can be identified on principal component scores and link to changes in the physical and chemical properties of agglomerates, thanks to the analysis of associated loading spectra (Fig. 2.25).

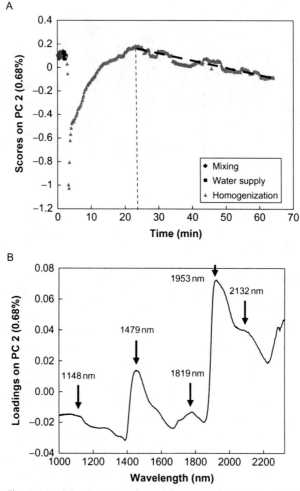

Figure 2.25 Changes in (A) NIR principal component scores PC$_2$ as a function of mixing time and (B) the respective spectra loadings calculated on raw spectra between 1000 and 2325 nm during semolina wet agglomeration. Arrows mark the most important peaks. Dotted lines show general tendencies. *Adapted from Mandato (2012).*

6. DIMENSIONAL ANALYSIS OF THE AGGLOMERATION PROCESSES

As previously stated, wet agglomeration processes are of primary importance in increasing the size of dried food materials and play a determinant role in the physical and functional properties of food powders. These processes encompass several steps, whose underlying mechanisms remain only partially understood. It is therefore difficult to determine rules of thumb regarding the influence of process conditions, which are necessary to raise the control of wet agglomeration processes to a higher level. In this section, we intended to propose a DA approach to fulfill this goal. After a brief reminder of the potential interests of DA in the food engineering field, the strategy envisaged to achieve the modeling of wet agglomeration processes by using a DA approach is expounded and the results obtained for the critical step of liquid atomization are detailed, before concluding by presenting the guidelines for a complete DA of wet agglomeration processes.

6.1. Benefits of using dimensional analysis in the modeling of wet agglomeration processes

The need for accurate modeling tools to link wet agglomeration process conditions to powder characteristics has clearly been evidenced in Section 1. In fact, the modeling of wet agglomeration processes by classical multiscale approaches (Dosta, Fries, Antonyuk, Heinrich, & Palzer, 2011) (e.g., computational fluid dynamics (Guo, Fletcher, & Langrish, 2004; Ho & Sommerfeld, 2002) and stochastic methods based on population balance equations (Dernedde, Peglow, & Tsotsas, 2011; Hapgood, Tan, & Chow, 2009; Terrazas-Verde, Peglow, & Tsotsas, 2011)) is an intricate task, mainly because of the following points:

- the complexity of agglomeration mechanisms (cf. Section 3) and the lack of physicochemical models to describe the formation of agglomerates (cf. Section 1),
- the broad scale range to take into account: atomic, microscopic, mesoscopic, and macroscopic scales for the determination of agglomerated powder characteristics (cf. Sections 1 and 2); and the laboratory, pilot, or industrial scales of the wet agglomeration processes (cf. Sections 1 and 3),
- the variety of wet agglomeration processes and equipments (cf. Section 5),
- the diversity of agricultural raw materials and agglomerated food products,
- the few agglomeration studies based on actual food materials (cf. Section 1).

Despite many recent breakthroughs in the understanding of the agglomeration mechanisms (Cuq et al., 2011; Fitzpatrick & Ahrné, 2005; Rondet, Delalonde, Ruiz, & Desfours, 2008; Rondet, Delalonde, et al., 2010) (cf. Sections 3–5), it is still difficult to establish process relationships allowing a quick prediction of the characteristics of agglomerated powders (e.g., mean particle diameter) as a function of the physicochemical properties of the inlet food material, agglomeration process design and equipment, and operating conditions. Moreover, empirical process relationships obtained by empirical approaches are valid only in a narrow range of process conditions for a given configuration of the wet agglomeration process. This precludes the possibility to apply process relationships, agglomeration process types or scales, and operating conditions directly to other food materials.

The use of DA is a way to overcome the numerous pitfalls in agglomeration processes modeling. Indeed, DA is a food engineering tool that possesses many qualities (Szirtes, 1998; Zlokarnik, 2001):

- DA is a well-established modeling approach that has been employed in the fluid dynamics and heat transfer fields for about a century: the π-theorem, on which DA is based, was developed about 100 years ago (Buckingham, 1914; Vaschy, 1892). Unfortunately, even though DA was successfully employed to scale-up cars, aircraft, vessels, heat exchangers, etc., this approach has only gained a modest acceptance in process engineering involving complex raw material.
- DA reduces the experimental framework required to establish process relationships by grouping the n dimensional parameters influencing the process in a set of q dimensionless ratios $(q < n)$ sufficient to define the system configuration.
- DA helps to identify the dimensionless numbers governing a given process, which often have a physical meaning that explains the mechanisms involved in the process.
- DA leads to building semiempirical relationships involving numerous dimensionless numbers under the influence of key inlet process parameters, and thus can be extrapolated to different process scales, designs, or operating conditions, and various food materials, provided that the underlying mechanisms are similar for the same order of magnitude of the dimensionless numbers within the investigated system configuration.

In this capacity, DA facilitates the design and scale-up of food processes. Finally, recent work has proved the efficiency of DA in modeling food processes involving complex raw material behaviors, such as gas–liquid mass transfer in an aerated tank (Hassan, Loubière, Legrand, & Delaplace, 2012), single kinetic reaction in a continuous reactor (Zlokarnik, 2001), protein denaturation, aggregation, and fouling in a plate heat exchanger (Petit, Méjean, et al., 2013; Petit, Six, Moreau, Ronse, & Delaplace, 2013), air inclusion in bakery foams (Delaplace, Coppenolle, Cheio, & Ducept, 2012), liquid atomization from mono- and bifluid nozzles (Mandato, Rondet, et al., 2012), and particle damage under stirring in glucose syrup suspensions (Bouvier, Moreau, Line, Fatah, & Delaplace, 2011).

Further, DA has recently been employed in the field of powder materials to model dry homogenization in a planetary mixer (André, Demeyre, Gatumel, Berthiaux, & Delaplace, 2012), mineral powder dissolution (André, Richard, Le Page, Jeantet, & Delaplace, 2012), and rehydration behavior of dairy powders (Jeantet, Schuck, Six, André, & Delaplace, 2010; Richard et al., 2012).

6.2. How to model wet agglomeration processes by DA

6.2.1 Implementation of a DA approach for food processes

DA of a food process to build a process relationship follows several steps. First, all the dimensional parameters impacting the food process need to be listed; as a prerequisite, the mechanisms involved in the process need to be well characterized. These key dimensional parameters are mainly related to process design (e.g., geometric parameters, characteristic lengths), process operating conditions (e.g., inlet temperatures, flow rates), physico-chemical characteristics of the inlet food material (e.g., concentration, density, viscosity), and physical constants (e.g., gravity acceleration, Boltzmann constant). Moreover, a target parameter, characteristic of the processed food material, should also be defined to build the process relationship.

Then, DA can be employed to turn the key parameters into dimensionless ratios by using the π-theorem. Indeed, this theorem states that a physical mechanism controlled by n dimensional parameters expressible in terms of p independent units can be completely described by a set of $q = n - p$ independent dimensionless numbers, formed by power products of the n key parameters (Szirtes, 1998; Zlokarnik, 2001). In other words, a process relationship linking the target parameter to the q independent dimensionless numbers exists and describes completely the considered food process. The use of the dimensional matrix method gives a dimensionless form to the target and key parameters (Delaplace, Maingonnat, Zaïd, & Ghnimi, 2009).

Lastly, model fitting tools are employed to find out the process relationship linking the target parameter in dimensionless form to the set of q independent dimensionless numbers, which gather the influence of the key process parameters.

6.2.2 Predominant influence of liquid atomization mechanism in wet agglomeration processes

The mechanisms taking part in the wet agglomeration processes are numerous and complex, as indicated in Sections 3 and 4 (Pietsch, 2005; Saad, 2011): wetting, nucleation, growth, consolidation, rupture, and stabilization

(Iveson et al., 2001). These wet agglomeration mechanisms involve many process parameters related to the physicochemical properties of the inlet food material (sprayed liquid and inlet fine powder particles), process design, and operating conditions (Jimenez-Munguia, 2007; Leuenberger, Betz, & Jones, 2006; Parikh, 2006).

Wet agglomeration can be seen as the result of surface wetting of powder particles and adhesion of particles owing to collisional or frictional forces (cf. Section 5). Thus, the characteristics of droplets formed by atomization (size, viscosity, surface tension, etc.) are of prime importance in the agglomeration process, as they will govern the wetting of the powder surface. Indeed, it has been evidenced in the literature that the mean droplet size is highly correlated to the final agglomerate size (Iveson et al., 2001; Jimenez-Munguia, 2007; Leuenberger et al., 2006; Parikh, 2006; Marmottant, 2011). This predominant influence of droplet size is also reflected in the dimensional spray flux (Lister, 2003) that is classically employed in wet granulation modeling to characterize the degree of dispersion of the liquid sprayed in the powder bed (cf. Section 3).

That is the reason why, prior to performing a complete DA of wet agglomeration processes, a DA of the atomization unit operation should be carried out to evaluate the influence of process parameters on the droplet size. This will be discussed in Section 6.3. Some guidelines to extend the DA of the atomization unit operation to the whole wet agglomeration process will be presented in Section 6.4.

6.3. Liquid atomization modeling by DA
6.3.1 Overview of the theory of liquid atomization: Identification of relevant dimensionless numbers

The mean size of droplets produced by liquid atomization units is controlled by numerous factors (Beau, 2006; Hede, Bach, & Jensen, 2008; Iveson et al., 2001; Jimenez-Munguia, 2007; Leuenberger et al., 2006; Mandato, Rondet, et al., 2012; Marmottant, 2011): for example, the physicochemical characteristics of the inlet liquid (viscosity, surface tension, and density) and air (viscosity, density, and humidity), operating conditions (liquid and air temperatures and flow rates), nozzle design (geometry, liquid and air orifice diameters), and characteristic lengths (distance to the nozzle).

The breakage of the liquid jet produced by a nozzle has been attributed to the frictional forces leading to Rayleigh instabilities, which are caused by the velocity difference between the injected liquid and the surrounding air (Hede et al., 2008; Mandato, Rondet, et al., 2012; Marmottant, 2011). In the case of monofluid nozzles, the surrounding air is almost static while the liquid velocity is high, which explains why an increase in liquid flow rate leads to smaller droplets. On the contrary, with bifluid nozzles, an increase in liquid flow rate has the same influence as a decrease in air pressure and leads to larger droplets, as it reduces the air–liquid velocity difference (Mandato, Rondet, et al., 2012). This shows that the air–liquid velocity difference is a crucial parameter in liquid atomization, which was expressed in recent work in dimensionless form by analogy with the well-known Weber number, classically used to evaluate drop breakage mechanisms (Marmottant, 2011; Weber, 1931). This dimensionless number, often called the "aerodynamic Weber number" and noted We_φ in this chapter, is given by Eq. (2.12) (Faragó & Chigier, 1992; Marmottant, 2011; Petit, Méjean, et al., 2013; Petit, Six, et al., 2013):

$$We_\varphi = \frac{\rho_A (u_A - u_L)^2 \varphi_L}{\sigma_L} \qquad [2.12]$$

with We_φ, aerodynamic Weber number; ρ_A, air density; u_A, air velocity at the nozzle outlet; u_L, liquid velocity at the nozzle outlet; φ_L, liquid jet diameter at the nozzle outlet, taken equal to the liquid orifice diameter; σ_L, liquid surface tension; all variables being expressed in SI units.

It is a useful dimensionless number that allows the description of the different regimes of liquid jet breakage, as it is highly correlated to experimental observations (Faragó & Chigier, 1992; Marmottant, 2011): for $We_\varphi < 25$, the liquid jet is broken in drops of similar size to the liquid orifice diameter; at $25 < We_\varphi < 70$, the liquid jet forms membranes that fragment in small droplets; and at $We_\varphi > 70$, the liquid jet is disrupted in liquid fibers and then in very small droplets, according to a fast atomization mechanism.

Another important dimensionless number has been mentioned in the literature for the characterization of liquid atomization in a given nozzle configuration: the air-to-liquid ratio $ALR = Q_A / Q_L$ (where Q_A and Q_L designate respectively the air and liquid mass flow rates in similar units), which is directly related to the ratio of fluid velocities. For instance, Hede et al. (2008) have shown that droplet size decreases with air-to-liquid ratio (ALR), whatever the bifluid nozzle geometry (Mandato, Rondet, et al., 2012).

The influence of the physicochemical properties of the sprayed liquid can be understood by the balance frictional forces and liquid resistance to deformation: the higher the liquid viscosity or the surface tension, the lower the influence of the frictional forces on the deformation of the liquid jet, leading to the formation of larger droplets.

Now that the main mechanisms acting in the liquid atomization processes and the significant dimensionless numbers have been identified from literature review, it is possible to implement a DA of liquid atomization.

6.3.2 DA of liquid atomization processes

A DA of the atomization process has to take into account the previously mentioned process parameters. Mandato, Rondet, et al. (2012) carried out a DA of atomization processes using mono- and bifluid nozzles of different geometrical characteristics in various process conditions (liquid flow rate, air pressure). This study also focused on the influence of the physicochemical characteristics (viscosity and surface tension) of the inlet liquid on droplet size measured by laser diffraction, by working with different ternary mixtures of water–ethanol–glycerol. They identified the most relevant process parameters in liquid atomization, namely, liquid flow rate and air pressure, and their conclusions regarding the influence of dimensional process parameters and liquid physicochemical properties were consistent with the atomization mechanism presented in Section 6.3.1.

The DA approach of Mandato, Rondet, et al. (2012) allowed the modeling of the droplet sizes obtained in a wide range of process conditions, physicochemical characteristics of the inlet liquid, and nozzle design with good accuracy (relative error between model and experimental results lower than 20%, Fig. 2.26).

The process relationship obtained by Mandato, Rondet, et al. (2012) for bifluid nozzles is given below (Eq. 2.13):

$$\frac{d_{50}}{L} = 1.66 \times 10^{22} \left(\frac{\rho_L}{L^{-1.5}g^{-0.5}\mu_A}\right)^{0.028} \left(\frac{\rho_A}{\rho_L}\right)^{0.030} \exp\left(-4 \times 10^{-5}\frac{\mu_L}{\mu_A}\right)$$

$$\left(\frac{\sigma_L}{L^{0.5}g^{0.5}\mu_A}\right)^{0.019} \left(\frac{S_A}{S_L}\right)^{0.94} \left(\frac{S_L}{L^2}\right)^{0.27} \left(\frac{\rho_A}{L^{-0.5}g^{0.5}\mu_A}\right)^{-2.2}$$

$$\exp\left(4 \times 10^{-2}\frac{u_L}{L^{0.5}g^{0.5}}\right) \tag{2.13}$$

with d_{50}, mean droplet size; L, measurement distance to the nozzle; ρ_L, liquid density; g, gravity acceleration; μ_A, air viscosity; ρ_A, air density; μ_L, liquid viscosity; σ_L, liquid surface tension; S_A, air orifice area; S_L, liquid orifice area; P_A, relative air pressure, u_L, liquid velocity at the nozzle outlet; all variables being expressed in SI units.

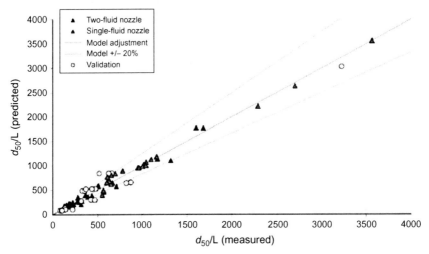

Figure 2.26 Predicted vs. experimental dimensionless droplet diameters for mono-
(gray triangles) and bifluid (black triangles) nozzles. White circles correspond to model
validation on additional measurements. *Extracted from Mandato, Cuq, and Ruiz (2012)
and Mandato, Rondet, et al. (2012).*

Moreover, this work highlights the respective roles of each process
parameter expressed in dimensionless form. In fact, comparing the absolute
values of correlation coefficients in Eq. (2.13) (process relationship) indicated that the respective roles of dimensionless process parameters decreased
in the following order (see Eq. 2.13): operating conditions (P_A and V_L) >
nozzle geometry (S_G and S_L) > physicochemical properties of inlet fluids
(ρ_L, ρ_G, μ_L, μ_G, and σ_L).

In a recent study (Petit, Méjean, et al., 2013; Petit, Six, et al., 2013), the
authors of this chapter applied the DA approach of Mandato, Rondet, et al.
(2012) both to milk concentrates and ternary mixtures of water–
ethanol–glycerol by using a different bifluid nozzle in a wide range of liquid
flow rates and air pressure. The AD approach developed by Mandato,
Rondet, et al. (2012) was extended by integrating the We_φ and ALR dimensionless numbers, whose significance toward liquid atomization has been
shown in Section 6.3.1. Our experimental and modeling results meet the conclusions of Mandato, Rondet, et al. (2012) regarding the predominant roles of
operating conditions over physicochemical properties of inlet liquid and air
(see Fig. 2.27). The decrease in the droplet size with We_φ or ALR increase was
also established in the following ranges: high We_φ ($304 < We_\varphi < 3108$)
corresponding to the violent "fiber disintegration regime" and moderate
ALR ($2.1 < ALR < 9.1$) values. These findings confirmed the influence of

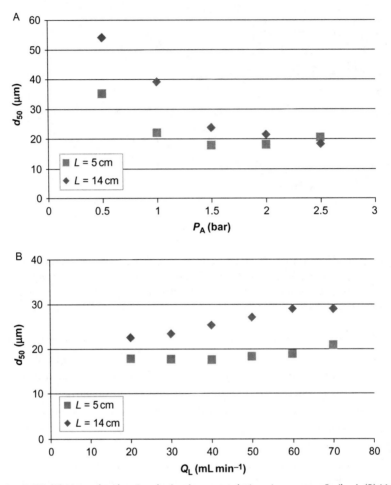

Figure 2.27 (A) Mean droplet size d_{50} (μm) versus relative air pressure P_A (bar). (B) Mean droplet size d_{50} (μm) versus milk concentrate flow rate Q_L (mL min⁻¹). Atomization of a 35% (w/w) milk concentrate with a bifluid nozzle. Laser diffraction measurement of mean droplet size at two distances to the nozzle outlet: 5 and 14 cm. (For color version of this figure, the reader is referred to the online version of this chapter.)

We$_{\varphi}$ and ALR dimensionless numbers reported in literature (already discussed in Section 6.3.1).

Our study put equal emphasis on another important variable in liquid atomization: the distance to the nozzle. We showed that the droplet size increased with the distance to the nozzle between 5 and 14 cm (see Fig. 2.27) whatever the operating conditions. This implies that the liquid atomization mechanism occurs rapidly at the nozzle outlet and then is followed by droplet coalescence owing to inertial effects. This parameter showed a considerable and complex influence in the liquid atomization process and was integrated in the DA approach by inserting it in each dimensionless number, as shown in the process relationship reported in Eq. (2.14)

$$
\frac{d_{50}}{L} = K(\text{ALR})^A \left(\frac{\rho_L}{L^{-1.5}g^{-0.5}\mu_A} \right)^B \exp\left(C\frac{\mu_L}{\mu_A} \right) \left(\frac{\sigma_L}{L^{0.5}g^{0.5}\mu_A} \right)^D
$$

$$
\text{We}_\varphi{}^E \left(\frac{\rho_A}{L^{-1.5}g^{-0.5}\mu_A} \right)^F \left(\frac{\varphi_L}{L} \right)^G
$$

[2.14]

with d_{50}, mean droplet size in volume; L, measurement distance to the nozzle; ALR, air-to-liquid ratio; ρ_L, liquid density; g, gravity acceleration; μ_A, air viscosity; μ_L, liquid viscosity; σ_L, liquid surface tension; We_φ, aerodynamic Weber number (calculated for the liquid jet); ρ_A, air density; φ_L, liquid orifice diameter; K, A, B, C, D, E, F, G, H, correlation coefficients (dimensionless); all variables being expressed in SI units.

This form of process relationship allows a good modeling of the droplet size of the investigated milk concentrates in the wide range of studied process parameters (about 15% accuracy, data not shown). The respective influence of each dimensionless number of this process relationship being still under investigation, it will not be discussed in this chapter.

6.4. Guidelines for a DA of wet agglomeration

The DA approach presented in Section 6.3 can be extended to achieve DA of wet agglomeration processes, such as powder agglomeration by recycling of the fine particles at the top of the chamber of a spray dryer. This will be studied at pilot scale by the authors of this chapter in a simpler configuration where the inlet fine powders are directly injected into the top of the drying chamber (Petit et al., 2012) (Fig. 2.28). The aim of this future experimental study is to extend the DA approach of the atomization process presented in Section 6.3.

Owing to the complexity of the wet agglomeration mechanisms involved in spray-drying processes and the lack of theoretical background describing the wet agglomeration mechanisms (already discussed in Sections 3–5), a blind DA of this wet agglomeration process is proposed.

Performing a DA of this wet agglomeration process requires the listing of additional relevant dimensional parameters that are necessary to describe the whole system, beyond the single spraying device. The relevant process parameters in the whole agglomeration system are constituted of the following (Table 2.3):

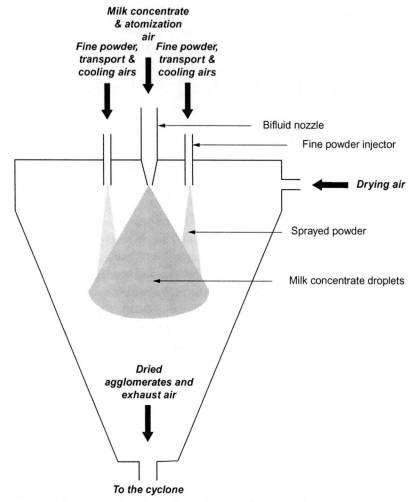

Figure 2.28 Scheme of the Minor drying pilot, supplied with a bifluid nozzle and modified for fine powder injection, which will be employed for wet agglomeration of skim milk powders (Petit et al., 2012). (For color version of this figure, the reader is referred to the online version of this chapter.)

- the key process parameters identified in the DA of liquid atomization (cf. Section 6.3),
- the physicochemical properties and operating conditions of additional processing airs (drying air, air for pneumatic conveying of the inlet fine particles—named transport air in Fig. 2.28, and cooling air): mainly, temperatures, humidity, and mass flow rate,

Table 2.3 Key process parameters in the investigated wet agglomeration system (Minor drying pilot modified for fine powder injection) and associated key dimensionless numbers

Process parameters	Process inputs				Process output
	Wet agglomeration system design	Inlet fine powder particles	Milk concentrate	Spraying, drying, cooling, and transport airs	Agglomerated powder
Geometry and configuration	Size and shape of the bifluid and powder nozzles, spray–drying pilot, and cyclone				
Physicochemical properties		Composition, size distribution, and shape	Concentration, density, viscosity, and surface tension	Density and viscosity	Composition, size distribution, and shape
Operating conditions		Velocity, humidity, temperature, and total injected mass	Flow rate and temperature	Flow rate, humidity, and temperature	
Key process parameters in this study	$\varnothing$ (no change in system configuration)	Q_p, d_p, m_p	μ_L, σ_L, u_L	T_{dry}, u_{spray}	d_{50}
Main dimensionless ratios	$\varnothing$	$\dfrac{Q_p}{L\mu_A}$, $\dfrac{d_p}{L}$, $\dfrac{m_p}{L^{1.5}g^{-0.5}\mu_A}$	$\dfrac{\mu_L}{\mu_A}$, $\dfrac{\sigma_L}{L^{0.5}g^{0.5}\mu_A}$, $\dfrac{u_L}{L^{0.5}g^{0.5}}$	$\dfrac{T_{dry}}{T_{amb}}$, $\dfrac{u_{spray}}{L^{0.5}g^{0.5}}$	$\dfrac{d_{50}}{L}$

$\dot{m}_p$, fine powder mass rate; d_p, fine powder mean size; m_p, total injected mass of fine powder particles; μ_L, liquid viscosity; σ_L, liquid surface tension; u_L, liquid velocity at the nozzle outlet; T_{dry}, drying air temperature; u_{spray}, spraying air velocity at the nozzle outlet; d_{50}, agglomerated powder mean size in volume; L, characteristic length (liquid nozzle orifice diameter for instance); g, gravity acceleration; μ_A, air viscosity at ambient temperature; T_{amb}, ambient temperature (20 °C). All variables being expressed in SI units in the dimensionless numbers (Petit et al., 2012).

- the physicochemical properties and operating conditions of the inlet fine powder particle properties: mainly, mass flow rate, mean particle diameter, and total injected mass during the process.

The target parameter can be chosen from the physicochemical properties (composition, size distribution, and shape) of the final agglomerated powder at the process outlet. In this DA approach, we focused on the mean agglomerate size in volume d_{50}.

The key process parameters and the target parameter were turned into dimensionless numbers by using the following set of independent quantities (L, g, T_{amb}, and μ_A), gathering the fundamental units (m, s, K, and kg) of the set of dimensional process parameters listed in Table 2.3. The identified key dimensionless numbers describe either the influence of physicochemical properties ($d_p/L, \mu_L/\mu_A, \sigma_L/L^{0.5}g^{0.5}\mu_A$) or the impact of operating conditions ($Q_p/L\mu_A, m_p/L^{1.5}g^{-0.5}\mu_A, u_L/L^{0.5}g^{0.5}, T_{dry}/T_{amb}, u_{spray}/L^{0.5}g^{0.5}$). The obtained dimensionless target parameter is in the form d_{50}/L. Process design parameters were excluded from this list of relevant dimensionless numbers, because system configuration will not be modified in the planned study.

This set of relevant dimensionless numbers could further be refined by integrating well-known dimensionless numbers having a clear physical meaning, such as those already identified by DA of liquid atomization (We$_\varphi$ and ALR, instead of $u_L/L^{0.5}g^{0.5}, u_{spray}/L^{0.5}g^{0.5}$), or the dimensionless spray flux Ψ_a (Lister, 2003). Moreover, this DA approach could be further extended to integrate the mechanism of fine powder recycling by adding the recycling rate and cyclone design characteristics in the list of relevant dimensional parameters. Also, the consideration of process geometrical features may improve the universality of process relationships obtained by DA.

In order to determine the roles of the established dimensionless numbers, pilot-scale experiments will soon be carried out with the Minor drying pilot supplied with a bifluid nozzle modified for fine powder injection represented in Fig. 2.28. For agglomeration purpose, skim milk powder particles with controlled characteristics (size, composition) will be injected in the spraying cone of skim milk concentrate (35%, w/w dried extract).

In conclusion, it has been shown in this section that DA is a powerful engineering tool, relatively easy to implement, which allows a quick identification of relevant process parameters, even for complex system configurations. The use of DA for modeling complex systems is recent, but this approach is increasingly gaining interest in food engineering. DA could be very useful in the field of food powders, as it throws light on the numerous and complex mechanisms involved in wet agglomeration processes and proposes a reasoned strategy to model the behavior of complex agricultural raw materials in agglomeration processes. Finally, as DA is a compulsory method when scaling up food processes, it should be highly considered in the modeling of wet agglomeration processes, which are multiscale by nature.

7. CONCLUSIONS

The multiplicity of mechanisms, which present complex interactions, makes the result of the food powder agglomeration still difficult to manage, in terms of qualities of the obtained agglomerates. A better determination of the relations coupling the operating variables with the physicochemical properties of agglomerates is still necessary to improve the understanding of the agglomeration process. This is more particularly true in the food industries, which still lacks theoretical foundations. The importance of industrial applications is considerable and explains the strong interest of the scientific community (essentially of the process engineering and food science) in powder agglomeration. One of the major stakes in these disciplines consists in proposing new criteria, functional diagrams, tools to monitor the processes, and new processes. The establishment of functional regime maps would allow predicting the flowing regime of particles thanks to pertinent dimensionless numbers. The diversity of the technologies and particle movements may generate a diversity of mechanisms (e.g., adhesion, coalescence, rupture, attrition, etc.) in which it would be advisable to determine the influences. To link the characteristic time of particle movements with those of agglomeration mechanisms (wetting, nucleation, growth, consolidation, break, and stabilization), it seems necessary to determine the characteristic times associated with the surface reactivity of the food particles, according to the local conditions of surface water content.

ACKNOWLEDGMENTS

The authors would like to thank the French National Research Agency (ANR) for its financial support through the scientific program "Reactive Powder" (ALIA 2008).

REFERENCES

Albert, R., Pfeifer, M. A., Barabasi, A. L., & Schiffer, P. (1999). Slow drag in a granular medium. *Physical Review Letters, 82,* 205–208.
Alcalà, M., Blanco, M., Bautista, M., & González, J. M. (2009). On-line monitoring of a granulation process by NIR spectroscopy. *Journal of Pharmaceutical Sciences, 99,* 336–345.

André, C., Demeyre, J. F., Gatumel, C., Berthiaux, H., & Delaplace, G. (2012). Dimensional analysis of a planetary mixer for homogenizing of free flowing powders: Mixing time and power consumption. *Chemical Engineering Journal, 198–199*, 371–378.

André, C., Richard, B., Le Page, J. F., Jeantet, R., & Delaplace, G. (2012). Influence of mixing system design and operating parameters on dissolution process. *Chemical Engineering and Technology, 35*, 247–254.

Aristov, Y. I., Gordeeva, L. G., Tokarev, M. M., Koptyug, I. V., Ilyina, L. Y., & Glaznev, I. S. (2002). H-1 NMR microimaging for studying the water transport in an adsorption heat pump. In: *Proceedings of the international sorption heat pump conference.*

Atterberg, A. (1911). Die plastizität der tone. *Internationale Mitteilungen Für Bodenkunde, 1,* 10–43.

Barbosa-Canovas, G. V., Ortega-Rivas, E., Juliano, P., & Yan, H. (2005). *Food powders: Physical properties, processing, and functionality.* Dordrecht: Kluwer Academic/Plenum Publishers.

Bardin, M., Knight, P. C., & Seville, J. P. K. (2004). On control of particle size distribution in granulation using high shear mixers. *Powder Technology, 140,* 169–175.

Barkouti, A., Rondet, E., Delalonde, M., & Ruiz, T. (2012). Influence of physicochemical binder properties on agglomeration of wheat powder in couscous grain. *Journal of Food Engineering, 111,* 234–240.

Barkouti, A., Turchiuli, C., Carcel, J., & Dumoulin, E. (2012). Milk powder agglomeration in fluidized bed: Influence of processing and product parameters on agglomerates growth and properties. In *5th International Symposium on Spray Drying Dairy Products, St Malo, France.*

Beau, P. A. (2006). Modélisation de l'atomisation d'un jet liquide—Application aux sprays diesel. PhD thesis. Université de Rouen, France.

Bertrand, D., & Dufour, D. (2000). *La Spectroscopie Infrarouge et ses Applications Analytiques.* Lavoisier, Paris: Tec et Doc.

Bouvier, L., Moreau, A., Line, A., Fatah, N., & Delaplace, G. (2011). Damage in agitated vessels of large visco-elastic particles dispersed in a highly viscous fluid. *Journal of Food Science, 76,* 384–391.

Buckingham, E. (1914). On physically similar systems: Illustrations of the use of dimensional equations. *Physical Review, 4,* 345–376.

Bunker, M., Zhang, J. X., Blanchard, R., & Roberts, C. J. (2011). Characterising the surface adhesive behavior of tablet tooling components by atomic force microscopy. *Drug Development and Industrial Pharmacy, 37,* 875–885.

Candelier, R., & Dauchot, O. (2009). Journey of an intruder through the fluidisation and jamming transitions of a dense granular media. *Physical Review E, 8,* 1.

Chen, X. D., & Ozkan, N. (2007). Stickiness, functionality, and microstructure of food powders. *Drying Technology, 25,* 959–969.

Chuy, L. E., & Labuza, T. P. (1994). Caking and stickiness of dairy-based food powders as related to glass transition. *Journal of Food Science, 59,* 43–46.

Conder, J. R., & Young, C. L. (1979). *Physicochemical measurement by gas chromatography.* Chichester, New York: John Wiley and Sons.

Condotta, R. (2005). Coulabilité des poudres cohésives: mesures aux faibles contraintes, granulaires humides et applications à une poudre industrielle. PhD Thesis. Institut Polytechnique National de Toulouse, France.

Coussy, O. (2011). *Mechanics and physics of porous solids.* New York: John Wiley & Sons, 296pp.

Cuq, B., Rondet, E., & Abecassis, J. (2011). Food powders engineering from knowhow to science: Industrial constraints, stakes and research opportunities. *Powder Technology, 208,* 244–251.

Dahlberg, C., Millqvist-Fureby, A., Schuleit, M., & Furo, I. (2010). Relationships between solid dispersion preparation process, particle size and drug release—An NMR and NMR microimaging study. *European Journal of Pharmaceutics and Biopharmaceutics, 76,* 311–319.

Delaplace, G., Coppenolle, P., Cheio, J., & Ducept, F. (2012). Influence of whip speed ratios on the inclusion of air into a bakery foam produced with a planetary mixer device. *Journal of Food Engineering, 108,* 532–540.

Delaplace, G., Maingonnat, J. F., Zaïd, I., & Ghnimi, S. (2009). Scale up of ohmic heating processing with food material. In E. Vorobiev, N. Lebovka, E. Van Hecke & J. L. Lanoisellé (Eds.), *Proceedings of international conference on Bio & Food Electrotechnologies (BFE2009), October 22–23, 2009 Compiègne* (pp. 101–107).

Derjaguin, B. V. (1989). *Theory of stability of colloids and thin films.* Berlin: Springer-Verlag, 276pp.

Dernedde, M., Peglow, M., & Tsotsas, E. (2011). Stochastic modelling of fluidized bed granulation: Influence of droplet pre-drying. *Chemical Engineering and Technology, 34,* 1177–1184.

Dosta, M., Fries, L., Antonyuk, S., Heinrich, S., & Palzer, S. (2011). Multiscale simulation of fluidised bed spray agglomeration. In *5th International Granulation Workshop, Lausanne, Suisse.*

Dowell, F. E., & Maghirang, E. (2002). Accuracy and feasibility of measuring characteristics of single kernels using near-infrared spectroscopy. In *"Novel raw materials, technologies, and products—New challenge for quality control," ICC conference 2002, Budapest, Hungary.*

Elversson, J., & Millqvist-Fureby, A. (2006). In situ coating—An approach for particle modification and encapsulation of proteins during spray-drying. *International Journal of Pharmaceutics, 323,* 52–63.

Ennis, B. J., Tardos, G. I., & Pfeffer, R. (1991). A microlevel-based characterization of granulation phenomena. *Powder Technology, 65,* 257–272.

Faldt, P., & Bergenstahl, B. (1996). Changes in surface composition of spray-dried food powders due to lactose crystallization. *Food Science and Technology, 29,* 438–446.

Faragó, Z., & Chigier, N. (1992). Morphological classification of disintegration of round liquid jets in a coaxial air stream. *Atomization and Sprays, 2,* 137–153.

Faure, A., York, P., & Rowe, R. C. (2001). Process control and scale-up of pharmaceutical wet granulation processes: A review. *European Journal of Pharmaceutics and Biopharmaceutics, 52,* 269–277.

Fitzpatrick, J. J., & Ahrné, L. (2005). Food powder handling and processing: Industry problems, knowledge barriers and research opportunities. *Chemical Engineering and Processing, 44,* 209–214.

Forny, L., Marabi, A., & Palzer, S. (2011). Wetting, disintegration and dissolution of agglomerated water soluble powders. *Powder Technology, 206,* 72–78.

Frake, P., Greenhalgh, D., Grierson, S. M., Hempenstall, J. M., & Rudd, D. R. (1997). Process control and end-point determination of a fluid bed granulation by application of near infra-red spectroscopy. *International Journal of Pharmaceutics, 151,* 75–80.

Fredlund, D. G., & Xing, A. (1994). Equations for soil-water characteristic curve. *Canadian Geotechnical Journal, 31,* 521–532.

Fyfe, K. N., Kravchuk, O., Le, T., Deeth, H. C., Nguyen, A. V., & Bhandari, B. (2011). Storage induced changes to high protein powders: Influence on surface properties and solubility. *Journal of the Science of Food and Agriculture, 91,* 2566–2575.

Fyfe, K., Kravchuk, O., Nguyen, A. V., Deeth, H., & Bhandari, B. (2011). Influence of dryer type on surface characteristics of milk powders. *Drying Technology, 29,* 758–769.

Gaiani, C., Ehrhardt, J. J., Scher, J., Hardy, J., Desobry, S., & Banon, S. (2006). Surface composition of dairy powders observed by X-ray photoelectron spectroscopy and effects on their rehydration properties. *Colloids and Surfaces. B, Biointerfaces, 49,* 71–78.

Gaiani, C., Morand, M., Sanchez, C., Tehrany, E. A., Jacquot, M., Schuck, P., et al. (2010). How surface composition of high milk proteins powders is influenced by spray-drying temperature. *Colloids and Surfaces. B, Biointerfaces, 75,* 377–384.

Gaiani, C., Scher, J., Ehrhardt, J. J., Linder, M., Schuck, P., Desobry, S., et al. (2007). Relationships between dairy powder surface composition and wetting properties during storage: Importance of residual lipids. *Journal of Agricultural and Food Chemistry, 55,* 6561–6567.

Gaiani, C., Schuck, P., Scher, J., Ehrhardt, J. J., Arab-Tehrany, E., Jacquot, M., et al. (2009). Native phosphocaseinate powder during storage: Lipids released onto the surface. *Journal of Food Engineering, 94,* 130–134.

Geng, J., & Behringer, R. P. (2005). Slow drag in 2D granular media. *Physical Review E, 71,* 011302.

Gravish, N., Umbanhowar, P. B., & Goldman, D. I. (2010). Force and flow transition in plowed granular media. *Physical Review Letters, 105,* 128301.

Gudehus, G. (2011). *Physical soil mechanics* (1st ed.). Berlin: Springer, 840pp.

Guo, B., Fletcher, D. F., & Langrish, T. A. G. (2004). Simulation of the agglomeration in a spray using Lagrangian particle tracking. *Applied Mathematical Modelling, 28,* 273–290.

Guyon, E. (2005). Disorders in granular matter. *Physica A: Statistical Mechanics and Its Applications, 357,* 150–158.

Hapgood, K. P., Litster, J. D., White, E. T., Mort, P., & Jones, G. (2004). Dimensionless spray flux in wet granulation: Monte Carlo simulations and experimental validation. *Powder Technology, 141,* 20–30.

Hapgood, K. P., Tan, M. X. L., & Chow, D. W. Y. (2009). A method to predict nuclei size distributions in models of wet granulation. *Advanced Powder Technology, 20,* 293–297.

Haque, E., Bhandari, B. R., Gidley, M. J., Deeth, H. C., & Whittaker, A. K. (2011). Ageing-induced solubility loss in milk protein concentrate powder: Effect of protein conformational modifications and interactions with water. *Journal of the Science of Food and Agriculture, 91,* 2576–2581.

Hartmann, M., & Palzer, S. (2011). Caking of amorphous powders—Material aspects, modelling and applications. *Powder Technology, 206,* 112–121.

Hassan, R., Loubière, K., Legrand, J., & Delaplace, G. (2012). A consistent dimensional analysis of gas-liquid mass transfer in a volume aerated stirred tank containing purely viscous fluids with shear-thinning properties. *Chemical Engineering Journal, 184,* 42–56.

Hede, P. D., Bach, P., & Jensen, A. D. (2008). Two-fluid spray atomisation and pneumatic nozzles for fluid bed coating/agglomeration purposes: A review. *Chemical Engineering Science, 63,* 3821–3842.

Heinrich, S., Blumschein, J., Henneberg, M., Ihlow, M., & Mörl, L. (2003). Study of dynamic multi-dimensional temperature and concentration distributions in liquid-sprayed fluidized beds. *Chemical Engineering Science, 58,* 5135–5160.

Ho, C. A., & Sommerfeld, M. (2002). Modelling of micro-particle agglomeration in turbulent flows. *Chemical Engineering Science, 57,* 3073–3084.

Iveson, S. M., Litster, J. D., Hapgood, K., & Ennis, B. J. (2001). Nucleation, growth and breakage phenomena in wet granulation processes: A review. *Powder Technology, 117,* 3–39.

Jafari, S. M., He, Y., & Bhandari, B. (2007). Role of powder particle size on the encapsulation efficiency of oils during spray drying. *Drying Technology, 25,* 1081–1089.

Jayasundera, M., Adhikari, B., Howes, T., & Aldred, P. (2011). Surface protein coverage and its implications on spray-drying of model sugar-rich foods: Solubility, powder production and characterization. *Food Chemistry, 128,* 1003–1016.

Jeantet, R., Schuck, P., Six, T., André, C., & Delaplace, G. (2010). The influence of stirring speed, temperature and solid concentration on the rehydration time of micellar casein powder. *Dairy Science & Technology, 90,* 225–236.

Jefferson, I., & Rogers, C. (1998). Liquid limit and the temperature sensitivity of clays. *Engineering Geology, 49,* 95–109.

Jimenez, T., Turchiuli, C., & Dumoulin, E. (2006). Particles agglomeration in a conical fluidized bed in relation with air temperature profiles. *Chemical Engineering Science, 61,* 5954–5961.

Jimenez-Munguia, M. T. (2007). Agglomération de particules par voie humide en lit fluidisé. PhD thesis. Ecole Nationale Supérieure des Industries Agricoles et Alimentaires, Vandoeuvre-lès-Nancy, France.

Jullien, R., & Botet, R. (1987). *Aggregation and fractal aggregates*. Singapore: World Scientific Publishing.

Kaddour, A. A., & Cuq, B. (2009). In line monitoring of wet agglomeration of wheat flour using near infrared spectroscopy. *Powder Technology, 190*, 10–18.

Kim, E. H. J., Chen, X. D., & Pearce, D. (2003). On the mechanisms of surface formation and the surface compositions of industrial milk powders. *Drying Technology, 21*, 265–278.

Kim, E. H. J., Chen, X. D., & Pearce, D. (2005a). Effect of surface composition on the flow-ability of industrial spray-dried dairy powders. *Colloids and Surfaces. B, Biointerfaces, 46*, 182–187.

Kim, E. H. J., Chen, X. D., & Pearce, D. (2005b). Melting characteristics of fat present on the surface of industrial spray-dried dairy powders. *Colloids and Surfaces. B, Biointerfaces, 42*, 1–8.

Kim, E. H. J., Chen, X. D., & Pearce, D. (2009a). Surface composition of industrial spray-dried milk powders. 1. Development of surface composition during manufacture. *Journal of Food Engineering, 94*, 163–168.

Kim, E. H. J., Chen, X. D., & Pearce, D. (2009b). Surface composition of industrial spray-dried milk powders. 2. Effects of spray drying conditions on the surface composition. *Journal of Food Engineering, 94*, 169–181.

Knight, P. C. (2001). Structuring agglomerated products for improved performance. *Powder Technology, 119*, 14–25.

Knight, P. C., Instone, T., Pearson, J. M. K., & Hounslow, M. J. (1998). An investigation into the kinetic of liquid distribution and growth in high shear mixer agglomeration. *Powder Technology, 97*, 246–257.

Larde, R., Bran, J., Jean, M., & Le Breton, J. M. (2009). Nanoscale characterization of powder materials by atom probe tomography. *Powder Technology, 208*, 260–265.

Le Meste, M., Lorient, D., & Simatos, D. (2002). *L'eau dans les aliments: Aspects fondamentaux. Signification dans les propriétés sensorielles des aliments et dans la conduite des procédés*. Paris: Editions Tec & Doc.

Leuenberger, H. (2001). New trends in the production of pharmaceutical granules: The classical batch concept and the problem of scale-up. *European Journal of Pharmaceutics and Biopharmaceutics, 52*, 279–288.

Leuenberger, H., Betz, G., & Jones, D. M. (2006). Scale-up in the field of granulation and drying. In M. Levin (Ed.), *Pharmaceutical process scale-up*. (2nd ed.). New York: Taylor & Francis (Chapter 8).

Lindberg, N. O., Leander, L., Wenngren, L., Helgesen, H., & Reenstierna, R. (1974). Granulation in a change can mixer. *Acta Pharmaceutica Suecica, 11*, 603–620.

Lister, J. D. (2003). Scale up of wet granulation processes: Science not art. *Powder Technology, 130*, 35–40.

Lister, J. D., Hapgood, K. P., Michaels, J. N., Sims, A., Roberts, M., & Kameneni, S. K. (2002). Scale-up of mixer granulators for effective liquid distribution. *Powder Technology, 124*, 272–280.

Litster, J. D., & Ennis, B. J. (2004). *The Science and Engineering of Granulation Processes*. Dordrecht: Kluwer Powder Technology Series.

Liu, A. J., & Nagel, S. R. (1998). Jamming is not just cool anymore. *Nature, 396*, 21.

Malamataris, S., & Kiortsis, S. (1997). Wettability parameters and deformational behaviour of powder-liquid mixes in the funicular agglomeration phase. *International Journal of Pharmaceutics, 154*, 9–17.

Mandato, S. (2012). Génie des procédés d'agglomération de poudres alimentaires: éléments de phénoménologie des apports d'eau et d'énergie mécanique. PhD thesis. Centre

International d'Etudes Supérieures en Sciences Agronomiques de Montpellier SupAgro, France.

Mandato, S., Cuq, B., & Ruiz, T. (2012). Experimental study of vertical stress profiles in a confined granular bed under static and dynamic conditions. *The European Physical Journal E, 35*, 56.

Mandato, S., Rondet, E., Delaplace, G., Barkout, A., Galet, L., Accart, P., et al. (2012). Liquids' atomization with two different nozzles: Modelling of the effects of some processing and formulation conditions by dimensional analysis. *Powder Technology, 224*, 323–330.

Marmottant, P. (2011). Atomisation d'un liquide par un courant gazeux. PhD thesis. Institut National Polytechnique de Grenoble, France.

Marshall, T. J., Holmes, J. W., & Rose, W. (1996). *Soil physics* (3rd ed.). Cambridge: Cambridge University Press, 472pp.

Matyas, E. L., & Radhakrishna, H. S. (1968). Volume change characteristics of partly saturated soils. *Geotechnique, 18*, 432–448.

McKenna, A. B., Lloyd, R. J., Munro, P. A., & Singh, H. (1999). Microstructure of whole milk powder and of insolubles detected by powder functional testing. *Scanning, 21*, 305–315.

Miller, B. S. (1979). A feasibility study on use of near infrared techniques. *Cereal Foods World, 24*, 88–89.

Montes, E. C., Dogan, N., Nelissen, R., Marabi, A., Ducasse, L., & Ricard, G. (2011). Effects of drying and agglomeration on the dissolution of multi-component food powders. *Chemical Engineering and Technology, 34*, 1159–1163.

Montes, E. C., Santamaria, N. A., Gumy, J. C., & Marchal, P. (2011). Moisture-induced caking of beverage powders. *Journal of the Science of Food and Agriculture, 91*, 2582–2586.

Muguruma, Y., Tanaka, T., & Tsuji, Y. (2000). Numerical simulation of particulate flow with liquid bridge between particles (simulation of centrifugal tumbling granulator). *Powder Technology, 109*, 49–57.

Murrieta-Pazos, I., Gaiani, C., Galet, L., Calvet, R., Cuq, B., & Scher, J. (2012). Food powders: Surface and form characterization revisited. *Journal of Food Engineering, 112*, 1–21.

Murrieta-Pazos, I., Gaiani, C., Galet, L., Cuq, B., Desobry, S., & Scher, J. (2011). Comparative study of particle structure evolution during water sorption: Skim and whole milk powders. *Colloids and Surfaces. B, Biointerfaces, 87*, 1–10.

Murrieta-Pazos, I., Gaiani, C., Galet, L., & Scher, J. (2012). Composition gradient from surface to core in dairy powders: Agglomeration effect. *Food Hydrocolloids, 26*, 149–158.

Nilpawar, A. M., Reynolds, G. K., Salman, A. D., & Hounslow, M. J. (2006). Surface velocity measurement in a high shear mixer. *Chemical Engineering Science, 61*, 4172–4178.

Osborne, B. G. (1999). Improved NIR prediction of flour starch damage. In L. O'Brien, A. B. Blakeney, A. S. Ross & C. W. Wrigley (Eds.), *Proceedings of 48th Australian cereal chemistry conference* (pp. 434–438). Melbourne: The Royal Australian Chemical Institute.

Osborne, B. G. (2000). Recent developments in NIR analysis of grains and grain products. *Cereal Foods World, 45*, 11–15.

Osborne, B. G. (2006). Utilisation de la spectroscopie proche infrarouge dans les industries céréalières. In D. Bertrand & E. Dufour (Eds.), *La spectroscopie infrarouge et ses applications analytiques* (pp. 505–535). Lavoisier, Paris: Tec & Doc.

Ozkan, N., Walisinghe, N., & Chen, X. D. (2002). Characterization of stickiness and cake formation in whole and skim milk powders. *Journal of Food Engineering, 55*, 293–303.

Palzer, S. (2011). Agglomeration of pharmaceutical, detergent, chemical and food powders—Similarities and differences of materials and processes. *Powder Technology, 206*, 2–17.

Palzer, S., Dubois, C., & Gianfrancesco, A. (2012). Generation of product structures during drying of food products. *Drying Technology, 30*, 97–105.

Palzer, S., & Fowler, M. (2010). Generation of specific product structures during spray drying of food. *Food Science and Technology*, *24*, 44–45.

Parikh, D. M. (2006). Batch size increase in fluid-bed granulation. In Michael Levin (Ed.), *Pharmaceutical process scale-up*. (2nd ed.). New York: Taylor & Francis (Chapter 9).

Perfetti, G., Van de Casteele, E., Rieger, B., Wildeboer, W. J., & Meesters, G. M. H. (2009). X-ray micro tomography and image analysis as complementary methods for morphological characterization and coating thickness measurement of coated particles. *Advanced Powder Technology*, *21*, 663–675.

Petit, J., Jeantet, R., Schuck, P., Janin, S., Méjean, S., Delaplace, G., et al. (2012). Multiscale modelling of dairy powder agglomeration mechanisms by a dimensional analysis approach. In *5th International Symposium on Spray Dried Dairy Products, Saint-Malo, France*.

Petit, J., Méjean, S., Accart, P., Galet, L., Delaplace, G., & Jeantet, R. (2013). Dimensional analysis of the atomisation of skim milk concentrates with a bi-fluid nozzle. *Journal of Food Engineering* (in preparation).

Petit, J., Six, T., Moreau, A., Ronse, G., & Delaplace, G. (2013). β-Lactoglobulin denaturation, aggregation, and fouling in a plate heat exchanger: Pilot-scale experiments and dimensional analysis. *Chemical Engineering Journal* (submitted for publication).

Pietsch, W. (2005). Agglomeration fundamentals. *Agglomeration in industry. Occurrence and applications: Vol. 1*. Weinheim: Wiley-VCH Verlag GmbH & Co. KGaA (Chapter 3).

Pouliquen, O., & Chevoir, F. (2002). Dense flows of dry granular material. *Compte rendus de l'Académie des Sciences de Paris, Physique*, *3*, 163–175.

Prime, D. C., Leaper, M. C., Jones, J. R., Richardson, D. J., Rielly, C. D., & Stapley, A. G. F. (2011). Caking behaviour of spray-dried powders—Using scanning probe microscopy to study nanoscale surface properties and material composition. *Chemical Engineering and Technology*, *34*, 1104–1108.

Prime, D. C., Stapley, A. G. F., Rielly, C. D., Jones, J. R., & Leaper, M. C. (2011). Analysis of powder caking in multicomponent powders using atomic force microscopy to examine particle properties. *Chemical Engineering and Technology*, *34*, 98–102.

Rahman, M. S. (2012). Applications of macro–micro region concept in the state diagram and critical temperature concepts in determining the food stability. *Food Chemistry*, *132*, 1679–1685.

Rajchenbach, J. (2003). Dense, rapid flows of inelastic grains under gravity. *Physical Review Letters*, *90*, 144302.

Rantanen, J., & Yliruusi, J. (1998). Determination of particle size in a fluidized bed granulator with a near-infrared spectroscopy (NIR) set-up. *Pharmacy and Pharmacology Communications*, *4*, 73–75.

Richard, B., Le Page, J. F., Schuck, P., André, C., Jeantet, R., & Delaplace, G. (2012). Towards a better control of dairy powder rehydration process. *International Dairy Journal*, In press.

Robert, P., Devaux, M. F., Mouhous, N., & Dufour, E. (1999). Monitoring the secondary structure of proteins by near-infrared spectroscopy. *Applied Spectroscopy*, *53*, 226–232.

Roman-Gutiérrrez, A. D. (2002). Propriétés d'hydratation des farines de blé: approches dynamiques et à l'équilibre. PhD Thesis. Centre International d'Etudes Supérieures en Sciences Agronomiques de Montpellier (Montpellier SupAgro).

Rondet, E., Delalonde, M., Ruiz, T., & Desfours, J. P. (2008). Hydrotextural and dimensional approach for characterising wet granular media agglomerated by kneading. *Chemical Engineering Research and Design*, *86*, 560–568.

Rondet, E., Delalonde, M., Ruiz, T., & Desfours, J. P. (2010). Fractal formation description of agglomeration in low shear mixer. *Chemical Engineering Journal*, *164*, 376–382.

Rondet, E., Denavaut, M., Mandato, S., Duri, A., Ruiz, T., & Cuq, B. (2012). Power consumption profile analysis during wet agglomeration process: Energy approach of wheat powder agglomeration. *Powder Technology*, *229*, 214–221.

Rondet, E., Ruiz, T., Delalonde, M., & Desfours, J. P. (2009). Hydrotextural description of an unsaturated humid granular media: Application to kneading and packing operations. *KONA Powder and Particle Journal, 27,* 174–185.

Rondet, E., Ruiz, T., Delalonde, M., Dupuy, C., & Desfours, J. P. (2010). Fractal law: A new tool for modeling agglomeration process. *International Journal of Chemical Reactor Engineering, 8,* A66.

Rouxhet, P. G., Misselyn-Bauduin, A. M., Ahimou, F., Genet, M. J., Adriaensen, Y., Desille, T., et al. (2008). XPS analysis of food products: Toward chemical functions and molecular compounds. *Surface and Interface Analysis, 40,* 718–724.

Ruiz, T., Rondet, E., Delalonde, M., & Desfours, J. P. (2011). Hydrotextural and consistency surfaces state of humid granular media. *Powder Technology, 208,* 409–416.

Ruiz-Cabrera, M. A., Foucat, L., Bonny, J. M., Renou, J. P., & Daudin, J. D. (2005). Assessment of water diffusivity in gelatine gel from moisture profiles. I—Non-destructive measurement of ID moisture profiles during drying from 2D nuclear magnetic resonance images. *Journal of Food Engineering, 68,* 209–219.

Russell, P., Diehl, B., Grinstead, H., & Zega, J. (2003). Quantifying liquid coverage and powder flux in high-shear granulators. *Powder Technology, 134,* 223–234.

Saad, M. (2011). Etude des mécanismes d'agglomération des poudres céréalières: Contribution des caractéristiques physiques et chimiques des particules à leur réactivité. Application à la fabrication du couscous. PhD thesis. Centre International d'Etudes Supérieures en Sciences Agronomiques de Montpellier SupAgro, France.

Saad, M., Barkouti, A., Rondet, E., Ruiz, T., & Cuq, B. (2011). Study of agglomeration mechanisms of food powders: Application to durum wheat semolina. *Powder Technology, 208,* 399–408.

Saad, M., Gaiani, C., Mullet, M., Scher, J., & Cuq, B. (2011). X-ray photoelectron spectroscopy for wheat powders: Measurement of surface chemical composition. *Journal of Agricultural and Food Chemistry, 59,* 1527–1540.

Saad, M., Gaiani, C., Scher, J., Cuq, B., Ehrhardt, J. J., & Desobry, S. (2009). Impact of re-grinding on hydration properties and surface composition of wheat flour. *Journal of Cereal Science, 49,* 134–140.

Saleh, K., & Guigon, P. (2009). Mise en œuvre des poudres. Techniques de granulation humide et liants. *Techniques de l'ingénieur, J2*(253), 1–14.

Sander, L. M. (1986). Fractal growth processes. *Nature, 322,* 789–793.

Schaafsma, S. H., Vonk, P., & Kossen, N. W. F. (2000). Fluid bed agglomeration with a narrow droplet size distribution. *International Journal of Pharmaceutics, 193,* 175–187.

Schoenherr, C., Haefele, T., Paulus, K., & Francese, G. (2009). Confocal Raman microscopy to probe content uniformity of a lipid based powder for inhalation: A quality by design approach. *European Journal of Pharmaceutical Sciences, 38,* 47–54.

Stauffer, D., & Aharony, A. (1994). *Introduction to percolation theory.* London: Taylor and Francis.

Szirtes, T. (1998). *Applied dimensional analysis and modelling.* New York: McGraw-Hill.

Szulc, K., & Lenart, A. (2010). Effect of agglomeration on flowability of baby food powders. *Journal of Food Science, 75,* 276–284.

Tardos, G. I., Khan, M. I., & Mort, P. R. (1997). Critical parameters and limiting conditions in binder granulation of fine powders. *Powder Technology, 94,* 245–258.

Terrazas-Verde, K., Peglow, M., & Tsotsas, E. (2011). Investigation of the kinetics of fluidized bed spray agglomeration based on stochastic methods. *AIChE, 57,* 3012–3026.

Travers, D. N., Rogerson, A. G., & Jones, T. M. (1975). A torque arm mixer for studying wet massing. *The Journal of Pharmacy and Pharmacology, 27*(Suppl.), 3.

Vaschy, A. (1892). Sur les lois de similitude en physique. *Annales Télégraphiques, 19,* 25–28.

Vignolles, M. L., Lopez, C., Ehrhardt, J. J., Lambert, J., Méjean, S., Jeantet, R., et al. (2009). Methods' combination to investigate the suprastructure, composition and properties of fat in fat-filled dairy powders. *Journal of Food Engineering, 94,* 154–162.

Vignolles, M. L., Lopez, C., Madec, M. N., Ehrhardt, J. J., Mejean, S., Schuck, P., et al. (2009). Fat properties during homogenization, spray-drying, and storage affect the physical properties of dairy powders. *Journal of Dairy Science*, *92*, 58–70.

Weber, C. (1931). Disintegration of liquid jets. *Zeitschrift für Angewandte Mathematik und Mechanik*, *11*, 136–159.

Wellm, A. B. (1997). Investigation of a high shear mixer/agglomerators. PhD thesis from The University of Birmingham, Royaume-Uni.

Wellner, N., Mills, C. E. N., Brownsey, G., Wilson, R. H., Brown, N., Freeman, J., et al. (2005). Changes in protein secondary structure during gluten deformation studied by dynamic Fourier transform infrared spectroscopy. *Biomacromolecules*, *6*, 255–261.

Yusof, Y. A., Smith, A. C., & Briscoe, B. J. (2005). Roll compaction of maize powder. *Chemical Engineering Science*, *60*, 3919–3931.

Zlokarnik, M. (2001). Scale-up of processes using material systems with variable physical properties. *Chemical and Biochemical Engineering Quarterly*, *15*, 43–47.

CHAPTER THREE

Dietary Strategies to Increase Satiety

Candida J. Rebello, Ann G. Liu, Frank L. Greenway,
Nikhil V. Dhurandhar[1]
Pennington Biomedical Research Center, Louisiana State University System, Baton Rouge, Louisiana, USA
[1]Corresponding author: e-mail address: Nikhil.Dhurandhar@pbrc.edu

Contents

Abstract

Obesity has a multifactorial etiology. Although obesity is widespread and associated with serious health hazards, its effective prevention and treatment have been challenging. Among the currently available treatment approaches, lifestyle modification to induce a negative energy balance holds a particularly larger appeal due to its wider reach and relative safety. However, long-term compliance with dietary modifications to reduce energy intake is not effective for the majority. The role of many individual nutrients, foods, and food groups in inducing satiety has been extensively studied. Based on this evidence, we have developed sample weight-loss meal plans that include multiple satiating foods, which may collectively augment the satiating properties of a meal. Compared to a typical American diet, these meal plans are considerably lower in energy density and probably more satiating. A diet that exploits the satiating properties of multiple foods may help increase long-term dietary compliance and consequentially enhance weight loss.

Advances in Food and Nutrition Research, Volume 69
ISSN 1043-4526
http://dx.doi.org/10.1016/B978-0-12-410540-9.00003-X

1. INTRODUCTION

Overweight and obese individuals are susceptible to several medical conditions that contribute to morbidity and mortality including diabetes and cardiovascular disease (Guh et al., 2009). According to the results of the National Health and Nutrition Examination Survey, 69.2% of adults in the United States and 16.9% of children and adolescents were overweight or obese in 2009–2010 (Flegal, Carroll, Kit, & Ogden, 2012; Ogden, Carroll, Kit, & Flegal, 2012). The prevalence of obesity among both males (35.5%) and females (35.8%) has not significantly changed in the two most recent years (2009–2010), as compared with the previous 6 years (Flegal et al., 2012). Although there is evidence for a leveling off in the prevalence of obesity (Rokholm, Baker, & Sorensen, 2010), the rise in health care costs attributable to overweight and obesity has been predicted to account for 16–18% of total U.S. health care costs (860.7–956.9 billion US dollars) by 2030 (Wang, Beydoun, Liang, Caballero, & Kumanyika, 2008). It is estimated that an obese individual in the United States is associated with approximately $1723 of additional medical spending per year (Tsai, Williamson, & Glick, 2011). The United States has a high per capita expenditure on health care; yet, life expectancy in the United States ranked 34th in the world in 2009 (WHO, 2011). Obesity has reduced longevity in several industrialized countries; however, the effects are more severe in the United States, which may be due to the unusually high rate of obesity in younger age groups and higher rates of severe obesity (Preston & Stokes, 2011).

Obesity, is a multifaceted problem with contributing factors that are undoubtedly complex and include genetics (Schwarz, Rigby, La Bounty, Shelmadine, & Bowden, 2011), endocrine function, behavioral patterns, and their environmental determinants (Gortmaker et al., 2011). Besides the well-known factors, several other putative contributors of obesity have been considered (McAllister et al., 2009). There is little argument about the health hazards of obesity. However, it is challenging to produce large and meaningful weight loss, and easy to regain weight. Although some research studies show weight loss and its benefits, there is an inability to successfully translate weight-loss strategies from a research setting to the general public. This issue was illustrated in a recent editorial as follows: "Consider a recent trial of weight loss and another of weight maintenance, which lasted an impressive two and three years, respectively (Sacks et al., 2009; Svetkey et al., 2008). These trials conducted by some of the world's foremost experts, on highly motivated individuals, who were closely monitored and supervised, resulted in about 4 kg reduction from about 100 kg starting body weight. While the results were statistically significant, their biological significance may be questioned. Will an approach that costs millions of dollars to produce less than 4% weight loss or maintenance, succeed in combating the global obesity epidemic in the free living population, where the facilities, the expertise of health care professionals, and the motivation of subjects is likely to be inferior to that in these studies?" (Dhurandhar, 2012).

Due to a lack of better alternatives, lifestyle modification, and obesity drugs or surgery in appropriately selected cases form the cornerstone of obesity management despite their limitations. Among these approaches, lifestyle modification to induce a negative energy balance holds a particularly larger appeal due to its wider reach and relative safety. Theoretically, negative energy balance could be achieved by increasing physical activity and reducing energy intake. However, recent studies demonstrate the inability of exercise by itself to make a significant contribution to weight loss (Thomas et al., 2012). This implies that the majority of the responsibility to induce a long-term negative energy balance that is adequate to reduce weight must be shouldered by dietary manipulations. However, modifying diet to reduce the long-term energy intake may work well for some responders but does not achieve adequate and lasting benefits for the majority. In most cases, the nonresponders or poor responders often outnumber the responders. It is time to recognize this limitation so that novel weight management approaches could be designed that are practical and effective.

Success of a weight-loss dietary regimen is closely related to compliance, which, in turn, is largely dependent on hunger, appetite, and satiety. As indicated by extensive research in the psychology, physiology, behavior, endocrinology, and pharmacology of hunger, appetite, and satiety, the significance of these factors is clearly well recognized. Much of the research about the effect of nutrients, foods, or food groups focuses on individual items. The role of many individual nutrients, foods, and food groups in inducing satiety is known, but not much attention has been focused on exploiting the combined effect of multiple satiating foods in a dietary regimen. Here, we have extensively and selectively reviewed the information available about the satiating and weight-loss properties of various common foods that may be relevant to a weight-loss regimen. The available evidence could be categorized into three groups: (a) effect on perceptions of satiety, (b) effect on actual food intake, and (c) weight loss as a result of a diet containing a satiating food. Based on this evidence, we have developed sample weight-loss meal plans that include multiple satiating foods, which may collectively augment the satiating properties of a meal.

1.1. Appetite and satiety

Appetite results from a convergence of several factors related to biology and the environment, referred to as the psychobiological system. It reflects the synchronous operation of events occurring on three levels: (1) psychological events and behavior, (2) peripheral physiology, and (3) the central nervous system (Blundell, 1999). However, apart from the desire to satisfy appetite sensations, in humans, sensory hedonics, sensory stimulation, tension reduction, social pressure, and boredom can also trigger an eating episode (de Graaf, Blom, Smeets, Stafleu, & Hendriks, 2004). Thus, a comprehensive definition of appetite would encompass the whole field of food intake, selection, motivation, and preference (Blundell et al., 2010).

Satiety is the process that inhibits further eating, causes a reduction in hunger, and an increase in fullness after a meal is eaten; whereas, the inhibitory processes that lead to the termination of a meal cause satiation. Though satiation and satiety are distinct concepts, they act together along with a host of other factors to determine eating behavior (Blundell et al., 2010). The satiety cascade proposed several years ago provides a framework for examining the processes (sensory, cognitive, post-ingestive, and post-absorptive) that mediate the satiating effect of foods (Blundell, 1999). A more recent modification to the satiety cascade includes the concepts of "liking" which is the pleasure derived from the orosensory stimulation of food and "wanting" which refers to the desire or motivation to actually engage in eating (Mela, 2006).

Eating behavior is controlled by metabolic factors that drive appetite and satiety, and sensory factors that influence food choice. In the brain, the sensory signals of food are associated with the metabolic consequences leading to a conditioning of eating and nutrition patterns. Cognitive factors such as an estimation of the satiating effect of foods and the timing of the next meal contribute to making eating a learned behavior (Blundell et al., 2010). It has been argued that food intake is controlled by an integrated set of signals at a momentary level. These signals are liable to change with the environment in which food is available, as each combination of items is consumed, and with every change in the physical and social context (Booth, 2008). By this argument, food intake tests and appetite-rating scales fall short in their assessment of satiety as they do not consider the changing influences over eating within, before, or after the test period, reactive responses, and the mental state of the individual at the precise moment that the quantitative judgment is being expressed (Booth, 2009).

Short-term control of eating is influenced by episodic signals that arise mainly from the gastrointestinal (GI) tract and are generated at regular intervals as food intake occurs. The peptides involved in GI signaling include cholecystokinin, glucagon-like peptide-1(GLP-1), peptide YY (PYY), and ghrelin. Long-term control of eating, also referred to as tonic signaling, reflects the metabolic state of adipose tissue. Tonic or enduring effects influence traits (stable predispositions), whereas episodic or transient control influences states (dispositions subject to rapid fluctuations) (Blundell et al., 2008). Psychometric tests such as the Eating Inventory (Stunkard & Messick, 1985) are used to identify traits that predispose individuals to restrained eating or a lack of control over eating or susceptibility to hunger. States reflect the drive to eat, are expressed as appetite ratings, and are measured using visual analog scales (VAS) (Blundell et al., 2008).

In a review of studies (de Graaf et al., 2004) that assessed satiety using subjective ratings of appetite, or by measuring actual energy intake, the vast majority of the studies showed that appetite ratings correlated with food intake in a standardized setting. Further, subjective appetite ratings and food intake were associated with changes in hormone concentrations (de Graaf et al., 2004). Subjective satiety responses usually correlate with the time of occurrence and magnitude of the effect of physiological processes such as increases in gastric volume and the absorption of nutrients (Blundell et al., 2010). Nevertheless, satiety claims do not need to be substantiated by physiologic data (Blundell, 2010). Appetite scores measured through VAS can be reproduced and are therefore feasible tools to measure appetite and satiety sensations (Blundell et al., 2010; Cardello, Schutz, Lesher, & Merrill, 2005; Flint, Raben, Blundell, & Astrup, 2000). Moreover, appetite and satiety have been found to be good determinates of food intake (Drapeau et al., 2005; Drapeau et al., 2007).

Macronutrient composition, energy density, and physical structure influence satiety (Blundell et al., 2010). Different amino acids, fatty acids, and carbohydrates have differing effects on the markers of appetite regulation (de Graaf et al., 2004). The differing effects imply that each nutrient interacts with the processes that mediate satiety in different ways (Blundell, 1999). Dietary protein may promote weight loss by increasing energy expenditure and stimulating satiety (Lejeune, Westerterp, Adam, Luscombe-Marsh, & Westerterp-Plantenga, 2006; Smeets, Soenen, Luscombe-Marsh, Ueland, & Westerterp-Plantenga, 2008; Westerterp-Plantenga, Nieuwenhuizen, Tome, Soenen, & Westerterp, 2009; Westerterp-Plantenga, Rolland, Wilson, & Westerterp, 1999). Carbohydrates that evade digestion, or which have a pronounced effect on glucose metabolism, have the potential to produce changes in appetite and affect satiety (van Dam & Seidell, 2007). Similarly, novel oils designed to prolong their presence in the GI tract can potentially produce beneficial effects on appetite and satiety (Halford & Harrold, 2012). The challenge lies in understanding which components of food interact optimally with the mediating processes to influence food intake. While the focus of this chapter is the effect of macronutrient composition, and energy density, of foods on appetite and satiety, the effect on body weight will also be explored.

2. DIETARY PROTEIN AND THE REGULATION OF FOOD INTAKE AND BODY WEIGHT

The most satiating macronutrient appears to be protein (Westerterp-Plantenga et al., 2009). High-protein diets with their potential to act on metabolic targets regulating body weight have become the subject of a body of research spanning 40 years (Halford & Harrold, 2012). Results from intervention studies suggest that an increase in the relative protein content of the diet reduces the risk of positive energy balance and the progress to weight gain (Clifton, Keogh, & Noakes, 2008; Weigle et al., 2005). Dietary protein may promote weight loss by increasing energy expenditure and by inducing satiety (Lejeune et al., 2006). In short-term studies (lasting for 24 h up to 5 days), evaluating subjective satiety sensations, high-protein diets have been shown to induce greater satiety than isoenergetic intakes of carbohydrate and fat (Lejeune et al., 2006; Westerterp-Plantenga et al., 1999).

Protein intake influences energy expenditure primarily through its effects on diet-induced thermogenesis. The thermic effect of nutrients is related to the adenosine triphosphate required for metabolism, storage, and oxidation (Westerterp-Plantenga et al., 2009). Three phosphate bonds are utilized for the incorporation of each amino acid into protein (Tome, Schwarz, Darcel, & Fromentin, 2009). The body is unable to store protein under conditions of high-protein intake and has to metabolize it which increases thermogenesis. Ultimately, resting metabolic rate also increases as a result of protein synthesis and protein turnover (Westerterp-Plantenga et al., 2009). In a study (Crovetti, Porrini, Santangelo, & Testolin, 1998) evaluating energy expenditure after consuming isoenergetic high-carbohydrate, high-protein, and high-fat meals, protein was found to be the most thermogenic nutrient and resulted in the greatest increase in sensations of fullness. Additionally, the study demonstrated a correlation between the thermic effect of food and eating behavior.

In a crossover trial (Lejeune et al., 2006), normal-weight women were prescribed an adequate protein diet consisting of 10% of energy from protein or a high-protein diet consisting of 30% of energy from protein. It was found that energy expenditure measured in a respiration chamber was greater during the higher protein intervention as a result of increased diet-induced thermogenesis and resting metabolic rate. In another respiratory chamber experiment, satiety and thermogenesis were shown to increase with a high-protein diet. Further, satiety was positively related to 24-h diet-induced thermogenesis (Westerterp-Plantenga et al., 1999).

Gluconeogenesis or the *de novo* synthesis of glucose from noncarbohydrate sources including amino acids is stimulated in the fed state by a high-protein diet (Veldhorst, Westerterp, & Westerterp-Plantenga, 2011). The energy expenditure that it requires as well as modulation of glucose homeostasis may have a role to play in protein-induced energy expenditure (Tome et al., 2009; Westerterp-Plantenga et al., 2009). Hepatic gluconeogenesis may be an alternative biochemical pathway to metabolize amino acids consumed in excess of requirements (Westerterp-Plantenga et al., 2009). However, not all amino acids can be used in gluconeogenesis; hence, the amino acid composition of the diet may have an effect on postprandial gluconeogenesis (Westerterp-Plantenga et al., 2009). In a comparison between a high-protein (30% of energy) and low-protein (12% of energy) diet, an increase in gluconeogenesis with the high-protein diet was demonstrated. However, there was no correlation between appetite ratings and gluconeogenesis (Veldhorst et al., 2011).

It has been hypothesized that protein-induced satiety is related to increased concentrations of the anorexigenic hormones GLP-1 and PYY and a decrease in the orexigenic hormone ghrelin (Batterham et al., 2006; Lejeune et al., 2006). In the respiratory chamber experiment (Lejeune et al., 2006), GLP-1 concentrations were measured nine times throughout the day, on the fourth day of consuming each of the diets. After dinner, GLP-1 concentrations were significantly higher on the high-protein diet when compared with the adequate protein diet. Energy expenditure, protein balance, and fat oxidation were also significantly higher on the high-protein diet in comparison with the adequate protein diet. Although ghrelin concentrations decreased, it could not be clearly attributed to the protein content of the diet, as the adequate protein diet with a relatively high-carbohydrate content also resulted in a decrease in ghrelin concentrations. Additionally, the increase in GLP-1 was related to the increase in satiety (Lejeune et al., 2006).

In a 3-week crossover trial, an effect of protein consumption on PYY concentrations was demonstrated with significantly higher plasma PYY and greater satiety responses to a high-protein meal in normal weight and obese individuals as compared with isoenergetic high-fat and high-carbohydrate meals (Batterham et al., 2006). However, in another study (Smeets et al., 2008), there were no differences in ghrelin and PYY responses between a high-protein (25% of energy) and average-protein (10% of energy) diet. GLP-1 response was in fact lower following the high-protein meal as compared with the average-protein (but higher carbohydrate) meal.

Specific amino acids may influence satiety by virtue of the fact that they are precursors for certain neurotransmitters involved in the regulation of appetite and body weight. Tryptophan is a precursor for the neurotransmitter serotonin. Tyrosine can be converted into the neurotransmitters dopamine and norepinephrine, and histadine can be converted into the neurotransmitter histamine. Each of these neurotransmitters has been linked with food intake regulation, although there is no direct evidence for their role in protein-induced satiety (Westerterp-Plantenga et al., 2009).

In a hypothesis propounded by Stock (1999), overeating a low- or high-protein diet is less metabolically efficient than overeating an average-protein diet (10–15% of energy). The metabolic inefficiency of a diet low or high in dietary protein stems from the energy cost involved in sparing lean body mass with a low-protein diet and building lean body mass with a high-protein diet (Westerterp-Plantenga et al., 2009). Energy intake required to build 1 kg of fat-free mass is much higher than the energy intake needed to build 1 kg of body weight with 60% fat mass and 40% fat-free mass (Stock, 1999). The data published in a recent study (Bray et al., 2012) demonstrated the metabolic inefficiency of low-protein diets. Weight gain in the low-protein group (3.16 kg) was less than in the normal (6.05 kg) and high-protein (6.51 kg) groups when the same number of extra calories was eaten over 56 days. Failure to increase lean body mass in the low-protein group accounted for their smaller weight gain.

By regulating food intake, diets high in protein have been shown to cause weight loss and affect body composition. The Recommended Dietary Allowance (RDA) for protein is 0.8 g/kg of body weight per day (NAP, 2005). In a meta-regression (Krieger, Sitren, Daniels, & Langkamp-Henken, 2006) classifying high-protein diets as >1.05 g/kg and low-protein diets as <1.05 g/kg of body weight per day, it was found that there was a greater association between fat-free mass retention and a high-protein diet as compared with a diet closer to the RDA (mean intake: 0.74 g/kg of body weight per day). Interestingly, in this meta-regression, while protein was found to be a good predictor of fat-free mass retention, it was not found to be a predictor of weight loss.

When subjects received a high-protein diet (30%, 50%, and 20% of energy from protein, carbohydrate, and fat, respectively), under controlled feeding conditions, satiety increased but body weight remained stable (Weigle et al., 2005). When the same diet was continued and subjects were advised to eat as much as they wanted (*ad libitum*), there was a sustained decrease in energy intake resulting in a weight loss, 76% of which comprised fat loss. In this study, protein was increased at the expense of the more energy-dense fat. In a cross-over trial (Blatt, Roe, & Rolls, 2011b), energy density was controlled, but the protein content of the diets was varied (10%, 15%, 20%, 25%, and 30% of energy). *Ad libitum* energy intake over 24 h did not vary significantly between the different protein levels. Additionally, satiety ratings were not significantly different across conditions.

Long-term (64 weeks) effects of a high-protein weight-loss diet (34%, 46%, and 20% of energy from protein, carbohydrate, and fat, respectively) were compared with an isoenergetic high-carbohydrate diet (17%, 64%, and 20% of energy from protein, carbohydrate, and fat, respectively) (Clifton et al., 2008). Weight loss was greater in participants who reported consuming a diet higher in protein. In a meta-analysis comparing high-protein–low-carbohydrate diets with high-carbohydrate–low-fat diets, weight loss was significantly greater with the high-protein diets in studies lasting up to 6 months, but the differences were not significant in studies spanning 12 months (Hession, Rolland, Kulkarni, Wise, & Broom, 2009).

Consumption of high-protein diets is not without some health risks. High levels of proteins can cause hypercalciuria and increase the rate of progression in renal dysfunction (Knight, Stampfer, Hankinson, Spiegelman, & Curhan, 2003). There appears to be evidence to indicate that high-protein diets enhance satiety and influence weight loss. However, the evidence that high-protein diets support weight maintenance over time is lacking, hence the need for long-term controlled trials on the safety and efficacy of high-protein diets.

2.1. Milk and milk products

Dairy products have been shown to induce satiety and reduce food intake (Dove et al., 2009; Gilbert et al., 2011; Harper, James, Flint, & Astrup, 2007). Additionally, dietary patterns that emphasize increased consumption of milk have been associated with the prevention of body weight gain (Drapeau et al., 2004). The regulatory effects on food intake and body weight associated with dairy intake have been attributed to the physiologic actions of its protein and calcium components (Gilbert et al., 2011; Lorenzen, Frederiksen, Hoppe, Hvid, & Astrup, 2012).

Casein and whey protein comprise 80% and 20%, respectively, of cow, sheep, goat, and buffalo milk (Luhovyy, Akhavan, & Anderson, 2007). Differences in the physical properties of casein and whey protein result in differing preabsorptive physiological effects. While whey protein remains soluble in the stomach and is emptied rapidly into the duodenum, casein forms a solid clot in the gastric acidic environment and is released slowly from the stomach (Boirie et al., 1997; Dangin, Boirie, Guillet, & Beaufrere, 2002). After whey protein ingestion, the plasma appearance of dietary amino acids is quick and high but of short duration. Further, it is associated with increased protein synthesis and oxidation with no change in protein catabolism. Casein ingestion results in a slower, more prolonged plasma appearance of amino acids. In comparison with whey, the metabolic response is different; it is accompanied by a considerable inhibition of protein breakdown, with a slight increase in synthesis and a moderate increase in oxidation (Boirie et al., 1997; Dangin et al., 2002). Both proteins, however, contribute to satiety following ingestion of dairy products (Luhovyy et al., 2007).

Milk proteins are a major source of a wide range of biologically active peptides. These peptides are inactive within the sequence of parent proteins and are released by enzymatic hydrolysis during digestion *in vivo*, or during processing of dairy products. Bioactive peptides in dairy products which have pharmacological similarities to opium are called opioid peptides. Opioid peptides are opioid receptor ligands with agonistic or antagonistic activities (Haque, Chand, & Kapila, 2009). Although endogenous opioid receptor agonists typically enhance feeding whereas antagonists inhibit feeding (Froetschel, 1996), their effects depend upon the macronutrient, individual food preference, and palatability of the food (Gosnell & Levine, 2009). Opioid peptides influence GI functions by affecting smooth muscles and reducing intestinal transit time (Haque et al., 2009). Opioid peptides known as β-casomorphins hydrolyzed from casein have been shown to delay gastric emptying and intestinal transit (Froetschel, 1996). The evidence linking opioid receptors to control of eating is supported largely by animal studies. From the evidence in human studies, it appears that opioid receptor antagonists have a small effect on hedonic taste preferences and short-term food intake. A number of clinical trials using naltrexone, an opioid receptor antagonist, have found no effects on body weight (Nathan & Bullmore, 2009).

As compared with other foods, whey proteins contain the highest concentrations of branched-chain amino acids (BCAA), especially leucine. Intracerebroventricular administration of either an amino acid mixture or leucine alone was shown to suppress 24-h food intake. It was determined that a rapamycin-dependent inhibition of Agouti-related protein gene expression contributed to the effect. Thus, increasing the amino acid concentrations within the brain suppressed food intake (Morrison, Xi, White, Ye, & Martin, 2007). BCAA, in general, and leucine, in particular, are first used for new protein synthesis, and diets that provide BCAA in excess of these requirements can support the intracellular leucine concentrations required to support other signaling pathways (Zemel, 2004).

Milk proteins stimulate the release of hormones involved in appetite regulation. Milk consumption modifies the insulinemic and glycemic response to carbohydrate-rich foods, with the whey fraction being the more efficient insulin secretagogue than casein (Nilsson, Stenberg, Frid, Holst, & Bjorck, 2004). Increased plasma concentrations of insulin have been associated with short-term satiety and decreased food intake (Samra, Wolever, & Anderson, 2007). The satiating effect of milk proteins has also been linked to the release of the anorexigenic hormones cholecystokinin, GLP-1, PYY, and suppression of the orexigenic hormone ghrelin (Luhovyy et al., 2007). Caseinomacropeptide, a digestion product of casein, collects in the whey fraction during cheese making. In rats, the glycosylated form of caseinomacropeptide, glycomacropeptide has been shown to increase pancreatic secretion (a marker of cholecystokinin) in a dose-dependent manner. Although glycomacropeptide may have potential as an appetite suppressant (Anderson & Moore, 2004), its effects on food intake remain uncertain (Luhovyy et al., 2007).

Whey protein has been found to be more satiating in some studies (Hall, Millward, Long, & Morgan, 2003; Veldhorst, Nieuwenhuizen, Hochstenbach-Waelen, van Vught, et al., 2009), while other studies have found no difference in the satiating effects of whey and casein protein (Bowen, Noakes, Trenerry, & Clifton, 2006; Lorenzen et al., 2012) or that casein protein leads to a greater satiating effect than whey proteins (Acheson et al., 2011). However, the effect of protein source is modulated by several factors including dose, form (solid or liquid), duration to the next meal, and the presence or absence of other macronutrients (Luhovyy et al., 2007). Nevertheless, energy intake was 9% lower after intake of milk than after intake of casein or whey (Lorenzen et al., 2012). Thus, complete milk proteins might elicit an intermediate yet optimal satiating effect, or other bioactive components in milk influence its satiating power.

The role of calcium intake on energy and fat balance has been explored, and studies that reviewed the available data provide some evidence for a beneficial effect of calcium on energy metabolism and the control of obesity (Major et al., 2008; Zemel, 2004). The "calcium appetite" theory (Tordoff,

2001) states that vertebrates develop a taste for and seek out minerals that they lack. Calcium-deficient individuals are more prone to increase their intake of foods high in fats and sugar since these nutrients are often found together with calcium (as in dairy foods). This preference for fat and sugar is the result of a learnt association between these nutrients and calcium.

Very few studies have assessed the impact of calcium consumption alone on the regulation of food intake. In women ($n = 13$) consuming <600 mg/day of calcium, supplementation with calcium + vitamin D reduced fat and total energy intake at an *ad libitum* food intake test. However, the sample size was small (Major, Alarie, Dore, & Tremblay, 2009). Subjects participating in a 6-month energy restriction program were assigned to either milk (1000 mg of calcium) or placebo (0 mg calcium) supplemented groups. Milk supplementation was accompanied by an increase in measured fullness that was significantly different from the decrease predicted by weight loss. Additionally, weight loss was found to induce stimulatory effects on appetite that were attenuated in the group receiving the milk supplementation (Gilbert et al., 2011). However, dairy supplementation increased protein intake; hence, it was difficult to distinguish between the effects of protein and calcium on food intake regulation.

In a crossover trial, four isoenergetic meals consisting of three meals with dairy products as the calcium source (low calcium [68 mg], medium calcium [350 mg], and high calcium [793 mg]) and one meal with a calcium supplement + calcium from dairy (total calcium: 850 mg) were compared. The four meals contained equivalent amounts of dairy protein. There was no significant effect of a high calcium intake from either dairy products or the supplement on appetite sensations, appetite hormones, or energy intake at a subsequent *ad libitum* meal (Lorenzen et al., 2007).

Satiety has been reported after consumption of dairy foods. Chocolate milk and a carbonated soft drink matched for energy density and energy were compared in a study to assess the satiety effects of the two beverages. Increased short-term satiety was observed after consumption of the chocolate milk when compared with the soft drink, but *ad libitum* energy consumption at lunch served 30 min later was not significantly different (Harper et al., 2007). The addition of 600 ml of skim milk to a fixed energy breakfast induced greater satiety than a fruit drink and reduced energy intake at a buffet sandwich meal 4 h later (Dove et al., 2009).

In a comparison of isoenergetic servings of yogurt, cheese, and milk, although there was no effect on energy intake, yogurt produced the greatest suppressive effect on appetite. Hunger ratings were 8%, 10%, and 24% lower after intake of yogurt as compared with cheese, milk, and water, respectively

(Dougkas, Minihane, Givens, Reynolds, & Yaqoob, 2012). However, in a comparison between isoenergetic, isovolumetric, servings of milk, high-fructose corn syrup sweetened, and sucrose-sweetened beverage preloads matched for energy density, no significant effect on satiety or food intake at a subsequent meal was observed (Soenen & Westerterp-Plantenga, 2007). Similarly, measures of satiety or food intake at a subsequent meal did not significantly differ between isoenergetic, isovolumetric preloads of orange juice, 1% milk, or a cola drink matched for energy density (Almiron-Roig & Drewnowski, 2003).

It is not only important that dairy products enhance satiety but it must be accompanied by a reciprocal compensation in energy intake through a reduction in nondairy foods. A cheesy snack containing a mixture of casein or whey + casein resulted in partial compensation in energy intake at an *ad libitum* lunch meal 1 h later and full compensation over the 24-h period following the preload. There was no difference in the satiety ratings between the two snacks (Potier et al., 2009). In a crossover trial, participants who were instructed to consume a high-dairy diet (three servings/day) or a low-dairy diet (one serving/day) for 7 days, separated by a 7-day wash-out period, increased their energy intake by 209 kcal during the high-dairy intervention. Further, there was no difference in appetite ratings between the two interventions (Hollis & Mattes, 2007b). However, a comparison of 200 kcal preloads consisting of semi-solid yogurt, liquid yogurt, dairy fruit beverage, and a fruit drink served 90 min prior to a meal resulted in increased satiety with the yogurt preloads as compared with the fruit beverages. Although the reduction in energy intake at the meal was not significant between the yogurt and fruit drinks, there was no energy compensation (Tsuchiya, Almiron-Roig, Lluch, Guyonnet, & Drewnowski, 2006).

There is not enough evidence to indicate that dietary calcium alone has regulatory effects on satiety and food intake. Dairy products include dairy proteins and other bioactive components that may have an effect on appetite control. The regulatory effects on satiety and food intake associated with increased dairy consumption have only been observed in the short term. Nevertheless, it demonstrates the need for investigating the various components of milk and milk products in the development of functional foods aimed at regulating appetite.

In a review of the epidemiological evidence on the effect of dairy consumption on body weight (Dougkas, Reynolds, Givens, Elwood, & Minihane, 2011), it was concluded that the data from cross-sectional studies supported an association between reduced adiposity and consumption of yogurt or milk, whereas cheese had the opposite effect. An inverse association was also observed between calcium intake from dairy and other sources with weight status in cross-sectional studies (Dougkas et al., 2011). However, in a systematic review of prospective cohort studies (Louie, Flood, Hector, Rangan, & Gill, 2011), it was concluded that the association between dairy consumption and weight status was inconsistent among children/adolescents as well as adults. Nevertheless, dairy consumption did not have an adverse effect on weight status, but the high heterogeneity of the studies and the inconsistent exposure and outcome measures precluded the conduct of a meta-analysis (Louie et al., 2011).

In two reviews (Dougkas et al., 2011; Lanou & Barnard, 2008) and a meta-analysis (Abargouei, Janghorbani, Salehi-Marzijarani, & Esmaillzadeh, 2012) of randomized controlled trials (RCTs) investigating the effects of dairy consumption on body weight, the authors concurred in their conclusion that body weight does not change due to increased consumption of dairy products in the context of diets without an energy restriction. Fat mass, lean body mass, and waist circumference were also unaffected by consumption of dairy products in the recommended amounts (three to four servings/day) when energy intake was not restricted (Abargouei et al., 2012). While one review of 11 studies investigating the effects of dairy or supplemental calcium intake on body weight in the context of energy-restricted diets found no effect of dairy or supplemental calcium intake on body weight (Lanou & Barnard, 2008), another review of five studies in which energy intake was restricted found inconsistent results (Dougkas et al., 2011). However, the conclusion from a meta-analysis of 11 studies imposing an energy restriction was that high-dairy energy-restricted diets may result in greater weight loss, fat loss, and higher reduction in waist circumference compared with energy-restricted diets that do not emphasize increased dairy consumption (Abargouei et al., 2012).

A more pronounced effect with equivalent calcium intakes from dairy products as opposed to the supplemental form in energy-restricted diets was also observed, suggesting that dairy components other than calcium may in part mediate the beneficial impact on body weight and composition (Dougkas et al., 2011). Thus, the inclusion of dairy products in the recommended amounts may prevent weight loss in weight maintenance diets, and high-dairy intake may facilitate weight loss in energy-restricted diets.

2.2. Meat, fish, and eggs

Dietary proteins provide amino acids responsible for major metabolic functions in the body. Protein quantity as well as quality influences its physiologic effects. Protein quality describes the characteristics of protein as they relate to the attainment of definite metabolic actions. The amino acid composition is the major determinant of protein quality (Millward, Layman, Tome, & Schaafsma, 2008). A complete protein provides all the essential amino acids (EAA) which include histidine, isoleucine, leucine, lysine, methionine, phenylalanine, threonine, tryptophan, and valine. Meat, poultry, fish, eggs, milk, cheese, and yogurt are considered complete proteins (Brown, 2008). Postprandial protein synthesis is enhanced when the composition of dietary protein matches the optimal amino acid needs of the body for protein synthesis. Thus, a well-balanced amino acid mixture as would be found in a complete protein would produce a higher thermogenic response than would an amino acid mixture of lower biological value (Westerterp-Plantenga et al., 2009).

The quality of a protein is influenced by the characteristics of the protein and the food matrix in which it is consumed, as well as the requirements of the individual consuming the food (Millward et al., 2008). The intake of protein over the whole day must provide as closely as possible the substrate needed for protein synthesis and any other biosynthetic pathways including a provision for sufficient signal amino acids (e.g., leucine) required for optimizing metabolism and stimulating anabolism (Millward et al., 2008; Young & Pellett, 1994). However, according to current methods of assessing protein quality, protein synthesis is limited by the available (digested and absorbed) EAA without considering regulatory amino acids (Millward et al., 2008).

While protein-induced satiety has been demonstrated in several studies (Harper et al., 2007; Lejeune et al., 2006; Smeets et al., 2008; Westerterp-Plantenga et al., 1999), there is less evidence for the role of protein from different sources on the regulation of appetite and food intake. Consumption of pork, beef, or chicken meals matched for energy and macronutrient content did not differ in their effects on satiety, food intake, or appetite hormones (cholecystokinin, PYY, ghrelin, and insulin) (Charlton et al., 2011).

Among protein-rich foods, fish was found to be the most satiating item, followed by beef, baked beans, eggs, cheese, and lentils in a study to assess the satiating capacity of several foods (Holt, Miller, Petocz, & Farmakalidis, 1995). In a comparison of beef, chicken, and fish meals containing 50 g protein, there was no difference in satiety between the chicken and beef meals; however, after consuming the fish meal, satiety increased as compared with the chicken and beef meals. The increase in the ratio of tryptophan to large neutral amino acids following consumption of the fish meal led the researchers to conclude that the neurotransmitter serotonin was one of the signals that induced satiety (Uhe, Collier, & O'Dea, 1992).

Liquid shakes containing protein from tuna, turkey, whey, or egg albumin that were matched for flavor, texture, and taste were compared.

Although the whey containing shake elicited the greatest satiety response, the tuna meal produced a significantly higher increase in satiety and a reduction in food intake than the turkey or egg shakes. The concentrations of protein, fat, and carbohydrate were matched in all the meals (Pal & Ellis, 2010). In another study, isoenergetic fish and beef meals matched for macronutrient composition, taste, and appearance were compared. The fish meal reduced energy intake at a subsequent meal compared with the beef meal; however, there were no significant differences in the appetite and satiety ratings (Borzoei, Neovius, Barkeling, Teixeira-Pinto, & Rossner, 2006).

The protein content of eggs is 35% of their total energy content (Pombo-Rodrigues, Calame, & Re, 2011). Isoenergetic egg (23% of energy from protein) and bagel (16% of energy from protein) breakfasts were compared. The egg breakfast significantly reduced the insulin, glucose, and ghrelin concentrations. Additionally, hunger was reduced and satisfaction increased after the egg breakfast as compared with the bagel breakfast, which resulted in a reduction in food intake at a subsequent meal (Ratliff et al., 2010). In another study comparing isoenergetic egg (18.3 g protein) and bagel (13.5 g protein) breakfasts matched for weight, consumption of the egg breakfast increased satiety and reduced energy intake at lunch. There was no compensation for the reduction in energy intake in the 24-h period following breakfast, as assessed by self-reported food intake (Vander Wal, Marth, Khosla, Jen, & Dhurandhar, 2005). Following consumption of three isoenergetic test lunches omelet, jacket potato, and chicken sandwich, the omelet meal was found to elicit a higher satiety response than the potato and chicken meals, but energy intake at dinner was not significantly different between the three conditions (Pombo-Rodrigues et al., 2011). In a comparison of isoenergetic egg and bagel breakfast meals matched for energy density, and consumed daily for 2 months, it was found that the egg breakfast did not promote weight loss. However, when subjects were placed on a reduced energy diet, eating the egg breakfast rather than the bagel breakfast resulted in greater weight loss (Vander Wal, Gupta, Khosla, & Dhurandhar, 2008).

Using NHANES data collected during the period from 1999 to 2004, it was determined that total meat consumption and "other meat products" (including frankfurter, sausage, organ meats, and food mixtures composed of meat poultry and fish) were positively associated with BMI and waist circumference. Half of the total meat consumption comprised other meat products. Of the remaining total meat consumption, red meat was consumed the most, followed by poultry and seafood (Wang & Beydoun, 2009). The European Prospective Investigation into Cancer and Nutrition

(EPIC) was a large-scale multicenter cohort study including subjects from 10 countries. Weight gain over 5 years among meat eaters, fish eaters, vegetarians (no meat or fish but including eggs and dairy products), and vegans (no foods of animal origins) from one arm of the EPIC study was assessed, and a statistically significant reduction in weight gain was found in vegans and fish eaters when compared with meat eaters. During the follow-up period, subjects who reduced the intake of meat had the lowest weight gain (Rosell, Appleby, Spencer, & Key, 2006).

Prospective studies investigating the effects of meat consumption on weight management using data from the EPIC study have yielded consistent results. Whereas an inverse association with red meat consumption and waist circumference was found among both males and females, a positive association between waist circumference and processed meat and poultry was found only in females. Fish and egg consumption was not associated with waist circumference in both groups (Halkjaer, Tjonneland, Overvad, & Sorensen, 2009). Using data from the EPIC study, an increase in meat intake of 250 g/day was determined to lead to a 2 kg weight gain in 5 years. In this cohort, red meat, poultry, and processed meat were positively associated with weight gain (Vergnaud et al., 2010). Data from six cohorts participating in the EPIC study showed that weight gain was positively associated with consumption of protein from red meat and processed meat, and poultry rather than fish and dairy (Halkjaer et al., 2011). However, fish consumption did not prevent an increase in waist circumference (Jakobsen et al., 2011). In another 7-year follow-up study, animal protein intake was found to be positively associated with overweight and obesity as assessed by BMI (Bujnowski et al., 2011).

Contrary results showing no significant associations between meat consumption and body weight or changes in adiposity measures over time were obtained in a 6-year follow-up study (Drapeau et al., 2004). Although protein intake was inversely associated with waist circumference, the effects were not significantly different between protein from animal and plant sources in subjects followed prospectively for 5 years (Halkjaer, Tjonneland, Thomsen, Overvad, & Sorensen, 2006).

A 2-year intervention trial comparing the effectiveness of a low-fat, energy-restricted, or Mediterranean diet determined that one of the leading predictors for 2-year successful weight loss was increased intake of meat and decreased intake of eggs (Canfi et al., 2011). In a RCT, participants who followed an energy-restricted diet containing lean or fatty fish had a greater weight loss than participants who followed an isoenergetic diet without fish

(Thorsdottir et al., 2007). A 12-week randomized controlled weight-loss trial among overweight women showed no significant difference in weight loss between subjects consuming an energy-restricted diet containing either lean beef or chicken (Melanson, Gootman, Myrdal, Kline, & Rippe, 2003).

RCTs investigating the effect of meat consumption on body weight are singularly lacking. The evidence associating meat intake with body weight is provided primarily by observational studies which precludes the establishment of a cause and effect relationship. There is a need for controlled trials investigating the effects of meat intake from various sources on appetite, satiety, food intake, and body weight.

2.3. Legumes

Legumes include alfalfa, clover, green beans and peas, peanuts, soy beans, dry beans, broad beans, dry peas, chickpeas, and lentils. Pulses are a type of legume exclusively harvested for the dry grain and include dry beans, chickpeas, lentils, and peas. Pulses are high in fiber which contributes to lowering the energy density and reducing the glycemic response, but they are also a good source of protein (McCrory, Hamaker, Lovejoy, & Eichelsdoerfer, 2010). A comparison of isoenergetic servings of chickpeas (16.1 g protein, 11.3 g fiber), lentils (18.3 g protein, 13.5 g fiber), navy beans (18.7 g protein, 16.6 g fiber), yellow peas (17.2 g protein, 9.1 g fiber), and a control of white bread (9.9 g protein, 2.8 g fiber) all served in a tomato-based sauce did not result in a significant difference in appetite and satiety ratings or food intake (Wong, Mollard, Zafar, Luhovyy, & Anderson, 2009). In the same study, canned beans and homemade beans were compared with glucose matched for carbohydrate content but differing in energy content. Satiety increased following consumption of homemade beans as compared with glucose, but canned beans consumption as compared with glucose reduced food intake at a pizza meal 2 h later (Wong et al., 2009).

Breads made by replacing 24.3% of wheat flour with chickpea flour (5 g fiber) or extruded chickpea flour (6 g fiber) were compared with wheat bread (3 g fiber) served as part of a breakfast meal. The meals were of equal food weight and had similar energy content. There were no significant differences in the satiety responses or energy intake at a buffet meal 2 h after the three breakfast meals, each containing one type of bread (Johnson, Thomas, & Hall, 2005). Pasta and tomato sauce meals matched for energy density and containing chickpeas, lentils, navy beans, yellow peas, or larger amounts of pasta and sauce were compared. The lentil treatment led to lower food intake compared to chickpeas and pasta with sauce, whereas navy beans led to lower intake compared only to chickpeas. However, there were no significant differences in food intake at a pizza meal 4 h later, although lentil treatment resulted in a significant decrease in cumulative intake (Mollard, Zykus, et al., 2011).

Isoenergetic preloads consisting of lentils and yellow peas but not chickpeas resulted in an increase in satiety and a reduction in food intake at a subsequent meal as compared with a "macaroni and cheese" meal (Mollard, Wong, Luhovyy, & Anderson, 2011). In another study assessing the effects of chickpea supplementation in the diet, participants were required to consume four 300 g cans of chick peas each week for 4 weeks. The increase in satiety at the end of 12 weeks was attributed to the increase in fiber intake (Murty, Pittaway, & Ball, 2010). Thus, the increase in satiety observed with pulse consumption may at least in part be attributed to the fiber content. However, the influence of pulses on satiety and food intake, particularly in comparison with other protein foods, and in the long term needs further substantiation through controlled studies.

The protein digestibility-corrected amino acid score is the currently approved method for protein quality assessment (Millward et al., 2008), and soy has been given a score of 1.0 which is the same as casein and egg protein. Thus, soy protein is considered a complete protein (Cope, Erdman, & Allison, 2008; Velasquez & Bhathena, 2007). In a study comparing isoenergetic breakfasts, matched for color, viscosity, and taste as assessed by VAS, satiety ratings were significantly higher after the breakfast meal containing 25% of energy from soy protein as compared with the meal containing 10% of energy from soy protein. However, food intake at a subsequent meal was not significantly different (Veldhorst, Nieuwenhuizen, Hochstenbach-Waelen, Westerterp, et al., 2009).

The satiating effects of three isoenergetic meals, which provided 50% of energy as whey, casein, or soy protein, were compared with a high-carbohydrate meal (95.5% energy from carbohydrate). The thermic effects of the protein-rich meals were significantly greater than the carbohydrate meal. Whey protein produced the highest thermogenic response, but casein and soy were more satiating than whey. However, satiety resulting from consumption of soy or casein proteins was not significantly different from the carbohydrate meal (Acheson et al., 2011). In a comparison of pretzels made with wheat flour and pretzels in which 27.3% of the wheat flour was replaced with soy ingredients, no significant effect on satiety was observed (Simmons, Miller, Clinton, & Vodovotz, 2011). In a comparison of soy protein, whey protein, and carbohydrate supplementation for

23 weeks, no significant effects on measures of satiety were observed (Baer et al., 2011). Although there appears to be some support for the satiety-enhancing effects of soy protein, a reduction in food intake remains to be established.

Using data from NHANES, an inverse association between high-pulse consumption and body weight was determined (Papanikolaou & Fulgoni, 2008). In a review (McCrory et al., 2010) of the effect of pulse consumption on weight management, five intervention trials that assessed the effectiveness of incorporating pulses in energy-restricted diets were identified (Abete, Parra, & Martinez, 2009; Hermsdorff, Zulet, Abete, & Martinez, 2011; Karlstrom et al., 1987; McCrory et al., 2008; Sichieri, 2002). In one study, a beneficial effect of legume consumption on weight loss at 1 month but not at 2 months was observed (Sichieri, 2002). Greater weight loss occurred with a medium intake (0.5 cup/day) as compared with a low intake (1 tablespoon/day) of pulses (McCrory et al., 2008). In a comparison of a high-legume diet (17% of energy from protein), a high-protein diet (30% of energy from meat, eggs, lean dairy), a fatty fish diet (17% energy from protein, mainly from fatty fish), and a control diet (17% of energy from protein, excluding legumes and fish), similar effects on weight loss were observed with the high-protein and high-legume diet. Both diets produced weight loss which was significantly greater than the control diet (Abete et al., 2009).

The consumption of legumes (four servings/week) within a low-energy diet resulted in a greater reduction in weight than a control diet that excluded legumes (Hermsdorff et al., 2011). However, in one study, no significant effect of legume consumption on body weight was observed (Karlstrom et al., 1987). The provision of pulses and whole grains for incorporation into the diet (during the first 6 months) also did not result in a significantly different weight loss after 6 or 18 months, as compared with a control group that was provided refined cereals and high-glycemic index (Gi) foods (Venn et al., 2010).

In a review (Cope et al., 2008) of the effect of soy protein on body weight, it was determined that epidemiological data provided inconsistent evidence for an inverse relationship between soy protein consumption and body weight. Clinical data did not indicate a clear advantage for soy protein over other sources of protein for weight and fat loss when consumed at isoenergetic levels. However, the comparator in most of the studies reviewed was milk (Cope et al., 2008). Subsequently, in a RCT assessing the effects of soy protein, whey protein, or carbohydrate supplementation for 23 weeks on measures of body weight and composition, there were no significant differences between soy and whey protein or soy and carbohydrate supplemented groups (Baer et al., 2011). Thus, the consumption of pulses and other legumes may enhance weight loss, but there does not appear to be sufficient evidence to support their role in weight management at this time.

2.4. Nuts

Nuts are fruits with one seed in which the ovary wall becomes hard at maturity. Common edible nuts include almonds, hazelnuts, walnuts, pistachios, pine nuts, cashews, pecans, macadamia, and Brazil nuts. Peanuts (groundnuts) are botanically a legume but are widely included in the nuts food group (Ros, 2010). Nuts are nutrient-dense foods rich in unsaturated fat and other bioactive compounds but are also high in energy (Ros, 2009). Nevertheless, evidence from epidemiological Albert, Gaziano, Willett, & Manson, 2002; Bes-Rastrollo et al., 2007; Fraser, Sabate, Beeson, & Strahan, 1992; Hu et al., 1998) as well as intervention trials (Natoli & McCoy, 2007; Rajaram & Sabate, 2006) suggest that nut consumption is not associated with weight gain. The mechanisms by which nuts impact body weight include increased satiety, increased energy expenditure, and incomplete digestion or absorption leading to increased fecal fat (Mattes, 2008; Rajaram & Sabate, 2006).

Nuts are energy-dense, and are high in fiber and protein, components of the diet associated with satiety (Mattes, Kris-Etherton, & Foster, 2008). The physical properties of nuts have also been implicated in their effects on satiety. Nuts require an effort at mastication, which may promote satiety (Slavin & Green, 2007). Nuts are high in fat and the prolonged mastication may influence satiety, since the metabolic effects of consumption of high-fat foods are modulated by oral exposure (Smeets & Westerterp-Plantenga, 2006). Nuts which have to be shelled have been shown to reduce energy intake as compared with nuts which have been shelled (Honselman et al., 2011). The act of shelling the nuts may slow the rate of consumption by permitting greater metabolic feedback during ingestion (Mattes, 2008) or may increase satiety by increasing perception of the quantity consumed (Kennedy-Hagan et al., 2011).

A diet containing walnuts was found to increase satiety ratings over a 3–4 day period as compared with a placebo diet (Brennan, Sweeney, Liu, & Mantzoros, 2010). However, no differences in satiety or energy intake were observed in a comparison of isoenergetic meals rich in polyunsaturated fats from walnuts, monounsaturated fat from olive oil, and saturated fat from dairy products (Casas-Agustench et al., 2009). Whole almonds (42.5 g) were found to increase fullness during the day when added to a breakfast meal matched for carbohydrate, protein, fat, and fiber when compared with meals containing almond flour, almond butter, almond oil, or no almonds (Mori, Considine, & Mattes, 2011).

It has been suggested that nut consumption leads to a spontaneous reduction in energy intake from other food sources during the day leading to an overall reduction in energy intake (Mattes, 2008). Almond supplementation for 6 months led to a 54–78% lack of compensation for the energy consumed from almonds (Fraser, Bennett, Jaceldo, & Sabate, 2002). From reported values of energy reduction, it has been estimated that approximately 65–75% of the energy provided from nuts is offset by lower energy intake at subsequent meals (Mattes et al., 2008).

An increase in resting energy expenditure (REE) following nut consumption has been observed. Regular consumption of peanuts for 19 weeks resulted in an 11% increase in REE when compared with the baseline measurement (Alper & Mattes, 2002). The high-unsaturated fat and protein content of nuts may increase fat oxidation and thereby influence diet-induced thermogenesis and REE (Mattes et al., 2008; Rajaram & Sabate, 2006). However, almond supplementation for 6 months did not increase REE (Fraser et al., 2002). In another study, diet-induced thermogenesis or REE did not change significantly after 10 weeks of almond supplementation (Hollis & Mattes, 2007a).

It is likely that limited fatty acid availability results from incomplete digestion or absorption. The parenchymal cell wall of nuts is resistant to microbial and enzymatic degradation. Thus, cells that are not ruptured as a result of insufficient mastication may pass through the GI tract without releasing the oils they contain (Mattes et al., 2008; Rajaram & Sabate, 2006). Using electron microscopy, it has been demonstrated that the cell walls of almonds remain intact in fecal samples, decreasing the bioaccessibility of intracellular fatty acids contained in the almonds and leading to a threefold increase in percent fecal fat excretion (Ellis et al., 2004). When participants chewed almonds 10 times, fecal fat losses significantly increased as compared with fecal fat excretion measured after they had chewed the almonds 25 or 40 times (Cassady, Hollis, Fulford, Considine, & Mattes, 2009). In a comparison of usual diets containing 70 g/day of whole peanuts, peanut oil, or peanut flour, fecal fat excretion was significantly higher with consumption of whole peanuts (Traoret et al., 2008). Further, other nutrients that contribute to energy are also less bioavailable which may cause further declines in energy intake (Mattes et al., 2008).

The Seguimiento Universidad de Navarra study is the only epidemiologic study that has prospectively examined the direct effect of nut consumption on body weight. The study included approximately 8800 adult men and women and found that those who ate nuts frequently ($\geq$two times/ week) had a 40% reduced risk of weight gain. During a follow-up period of 28 months, frequent nut consumers gained 350 g less weight than did those who did not eat nuts (Bes-Rastrollo et al., 2007). In a recent review of the epidemiologic evidence on the effect of nut consumption on weight gain and obesity, it was concluded that nut consumption up to four servings/ week does not lead to any appreciable weight gain in the long term (Martinez-Gonzalez & Bes-Rastrollo, 2011).

The impact on body weight of consumption of almonds, peanuts, pistachios, and walnuts has been examined in intervention trials. Six months supplementation of almonds averaging (320 kcal/day) resulted in a weight gain of 0.4 kg. The predicted weight gain as a result of the extra energy intake from the almonds was 6.4 kg (Fraser et al., 2002). In another study, subjects consuming an almond-enriched (84 g/day), low-calorie diet for 24 weeks lost 62% more weight and showed significant improvements in measures of body composition than subjects assigned to a low-calorie complex carbohydrate diet (Wien, Sabate, Ikle, Cole, & Kandeel, 2003). In contrast, reduced energy almond-enriched (56 g/day) or nut-free diets resulted in a greater reduction in weight loss in the nut-free diet condition at 6 months in overweight and obese individuals. There was, however, no difference in weight loss between the diets at 18 months (Foster et al., 2012). It has also been reported that almond supplementation (approximately 344 kcal) for 10 weeks caused no significant change in body weight or body composition (Hollis & Mattes, 2007a).

In other studies, peanut supplementation for 6 months increased energy intake; however, the actual weight gain was less than the predicted weight gain as a result of partial compensation for energy provided by peanuts (Alper & Mattes, 2002). A low-calorie weight-loss trial comparing the effects of pistachios or pretzels served as an afternoon snack for 12 weeks found a significantly greater reduction in the BMI of participants consuming pistachios (Li et al., 2010). Daily consumption of either 42 g or 70 g of pistachios for 12 weeks did not lead to weight gain or an increase in waist-to-hip ratio in Chinese subjects with metabolic syndrome (Wang, Li, Liu, Lv, & Yang, 2012). Participants following their usual diet were provided with walnuts corresponding to 12% of their daily energy intake (28–56 g). At the end of 6 months, the theoretical weight gain was predicted to be 3.1 kg. Although daily energy intake increased by 133 kcal, the weight gain was only 0.4 kg, neither of which were significant changes (Sabate, Cordero-Macintyre, Siapco, Torabian, & Haddad, 2005).

Thus, increased satiety with nut consumption, the displacement of foods from the habitual diet, and increased fecal fat excretion appear to be plausible mechanisms by which nut consumption does not adversely influence body weight despite the fact that nuts are energy dense. However, the effects of nut consumption on diet-induced thermogenesis and REE need further substantiation. Inclusion of nuts in energy-restricted diets may help adherence to the diet, facilitate weight loss, and improve measures of body composition.

3. CARBOHYDRATES AND THE REGULATION OF FOOD INTAKE AND BODY WEIGHT

Carbohydrates influence satiety and food intake through several mechanisms related to their hormonal effects, intrinsic properties, and intestinal fermentation (Beck, Tapsell, Batterham, Tosh, & Huang, 2009; Beck, Tosh, Batterham, Tapsell, & Huang, 2009; Greenway et al., 2007; Hamedani, Akhavan, Abou Samra, & Anderson, 2009; Ludwig, 2002; Vitaglione, Lumaga, Stanzione, Scalfi, & Fogliano, 2009). The hormonal effects of carbohydrates on satiety are mediated by insulin (Ludwig, 2002) and GI hormones (Beck, Tapsell, et al., 2009; Vitaglione et al., 2009), whereas the intrinsic properties relate to the bulking and viscosity effects of dietary fiber (Beck, Tosh, et al., 2009; Hamedani et al., 2009). Carbohydrates that elude small intestinal digestion enter the large bowel and are fermented by colonic microorganisms into short-chain fatty acids which have been shown to enhance satiety (Greenway et al., 2007). The manipulation of the carbohydrate content of the diet has been explored as a means of regulating body weight.

3.1. Glycemic index

The Gi was developed as a method for classifying the carbohydrates in different foods according to their post-ingestion glycemic effect (Jenkins et al., 1981). It is defined as the incremental area under the curve (AUC) for the blood glucose response after consumption of a 50 g carbohydrate portion of a test food expressed as a percent of the response to an equivalent carbohydrate amount from a reference food ingested by the same subject (Wolever, Jenkins, Jenkins, & Josse, 1991), with glucose or white bread as the reference food. Based on a commonly accepted classification system, foods are categorized as low Gi (<55) or high Gi (>70) (Venn & Green, 2007) For a meal, and by implication, a diet, the Gi of individual food items may be used to predict the Gi of a mixed meal (Wolever et al., 1991). However, the quantity and source of carbohydrate are important components affecting the glycemic response (Venn & Green, 2007).

The concept of glycemic load (GL) has been developed to take into account the amount of carbohydrate consumed. The GL value of a food may be determined indirectly as a product of the Gi of a food and the amount of available carbohydrate (carbohydrate that is absorbed by the small intestine and used in metabolism (Livesey, 2005)) in the portion of food consumed. The glycemic equivalence is a method of directly determining the GL. It involves constructing a standard curve for each subject based on the AUC for glucose calculated for a range of doses of the reference food measured on different days. The AUC for a given food consumed at any portion size is compared to the individual's standard curve. Theoretically, it is the amount of glucose that would produce the same blood glucose AUC as that particular portion size of food consumed. It is a time consuming and costly method which in any event agrees well with the GL measured indirectly (Venn & Green, 2007).

The potential physiological mechanisms relating Gi to the regulation of food intake are based on the postprandial metabolic environment precipitated by hyperglycemia and hyperinsulinemia. It has been suggested that a high-GL meal elicits high insulin and low glucagon responses that promotes uptake of glucose in the muscle, liver, and fat tissue. Hepatic glucose production is thereby restrained and lipolysis is inhibited. Thus, denial of full access to the two major metabolic fuels in the post-absorptive state may lead to a quicker hunger response and overeating, as the body attempts to restore the concentration of metabolic fuels to normal (Ludwig, 2002). Low-Gi foods are characterized by a slow rate of digestion and absorption, thereby eliciting a low glycemic response (Wolever et al., 1991).

Increased short-term satiety with low-Gi foods or meals as compared with high-Gi foods or meals has been demonstrated in a vast majority of the studies investigating the effects of the diets based on the Gi (Livesey, 2005). A systematic review of the effect of low-Gi diets on satiety and body weight in the long term (several days or weeks duration) found inconsistent results (Bornet, Jardy-Gennetier, Jacquet, & Stowell, 2007). More recently, a 10-week parallel study (Krog-Mikkelsen et al., 2011) to investigate the effects of low-Gi or high-Gi diets found no differences in postprandial plasma leptin and ghrelin to support the satiating effect of either Gi diet. Subjective appetite sensations, energy expenditure, or substrate oxidation were also not significantly different.

The association between Gi or GL and obesity has been the focus of numerous studies, and the subject of controversy, yet many best-selling diet books including the South Beach diet, the Zone diet, and the New Glucose Revolution advocate consumption of a low-Gi diet. Although some epidemiological studies indicate that increasing the Gi of a diet is associated with an increase in BMI, a collective analysis of the data does not support the view that a high-Gi diet is predictive of a higher BMI when compared with a lower Gi diet. The majority of epidemiological evidence also does not support the belief that high-GL diets are predictive of adiposity (Gaesser, 2007).

A meta-analysis of six RCTs (Thomas, Elliott, & Baur, 2007) that compared low-Gi/GL diets with either high-Gi/GL or low-fat diets in overweight or obese individuals concluded that low-Gi or -GL diets resulted in a decrease in body mass by 1.1 kg, fat mass by 1.1 kg, and body mass index by 1.3 kg/m². Another meta-analysis of 23 studies (Livesey, Taylor, Hulshof, & Howlett, 2008) that measured weight loss following low-Gi/GL diets showed that a reduction in body weight occurred with a reduction in dietary GL when food intake was either *ad libitum* or relatively uncontrolled. The beneficial effect, however, was not observed in studies where the food intake was controlled. Low-GL diets prescribed *ad libitum* have also been shown to be effective in promoting weight loss in overweight adolescents in comparison with reduced-fat diets (Ebbeling, Leidig, Sinclair, Hangen, & Ludwig, 2003).

A RCT (Maki, Rains, Kaden, Raneri, & Davidson, 2007) to evaluate the effects of an *ad libitum* reduced-GL diet on body weight and body composition in overweight or obese adults found that subjects assigned to the reduced GL diet lost significantly more weight than those assigned to a low-fat diet at 12 weeks. However, there was no significant difference in body weight between the groups at 36 weeks despite continuation of the diet in the weight maintenance phase following weight loss. In the CALERIE trial, a 1-year RCT designed to examine the effects of high- or low-GL, calorie-restricted diets on weight and fat loss in overweight women, there was no significant change in body weight, body fat, resting metabolic rate, hunger, and satiety between the groups (Das et al., 2007). Thus, it appears that reduced energy intake with low-GL diets may not persist in the long term. Analysis of 4-day food records, in a randomized crossover intervention (Aston, Stokes, & Jebb, 2008), comparing two diets differing in the Gi, suggested that 19 overweight or obese women consumed comparable amounts of energy and macronutrients on both diets and more dietary fiber on the low-Gi diet. There was, however, no significant difference in body weight, body composition, and waist circumference between the two intervention periods.

The clinical relevance of diets based on the Gi or GL remains unclear. The inconsistencies in the data appear to be largely responsible for the debate surrounding the concept. The Gi is influenced by the nature of the starch, the physical form, the amount of fiber, fat, and protein, as well as the cooking times and methods (Thorne, Thompson, & Jenkins, 1983). Other dietary factors influencing food digestibility, GI motility, or insulin secretion also determine the Gi of a food (Ludwig, 2003). The GI relates to a food and not the individual; therefore, there exists the possibility of intra- and inter-individual variances in the Gi and GL. Even in repeated experiments of the same food under standardized conditions, a seemingly inexplicable variation occurs in the glycemic response (Venn & Green, 2007). Such variability makes following a low-Gi diet complex.

The question remains as to whether the individual glycemic indices of foods can be summed to reliably derive the Gi of the meal. While some researchers contend that the Gi predicts the glycemic response of foods eaten as part of a mixed meal, others have shown no association between the calculated Gi and the measured Gi of the meal as a whole (Venn & Green, 2007). Nevertheless, the most important point of issue is the practicality of recommending Gi diets.

None of the studies reviewed in this chapter have suggested any adverse effects from consuming a low-Gi diet. In general, nonstarchy vegetables, fruits, legumes, and minimally processed grain products have a low Gi (Ludwig, 2003). A diet that includes these food groups meets the recommendations of the Dietary Guidelines for Americans, 2010 and could contribute to increasing the fiber and lowering the energy density of a diet. Central to the concept of low Gi is the emphasis on carbohydrate quality, which presents less of a challenge to glucose homeostasis than high-Gi diets. Low-Gi and -GL diets have been recommended in the treatment of diabetes and prevention of chronic diseases including diabetes, cancer, and cardiovascular disease (Jenkins et al., 2002). Moreover, benefits have been shown from *ad libitum* consumption of low-Gi diets (Ebbeling et al., 2003; Livesey et al., 2008) which suggests that these diets may be less restrictive and more acceptable as a dietary approach to the regulation of body weight. Nevertheless, the data do not present an unequivocal association between low-Gi and -GL diets, and reductions in body weight.

3.2. Dietary fiber

The World Health Organization and the Food and Agriculture Organization consider dietary fiber to be a polysaccharide with 10 or more monomeric units which is not hydrolyzed by endogenous hormones in the small intestine (Lattimer & Haub, 2010). Dietary fiber may be classified into soluble and insoluble fiber on the basis of water solubility. Colonic fermentation of soluble fiber yields short-chain fatty acids. Insoluble fiber generally has low fermentability, but it has water-attracting properties that promote fecal bulk (Papathanasopoulos & Camilleri, 2010).

Several mechanisms have been proposed to explain the effects of dietary fiber on the regulation of appetite and satiety. (1) Dietary fiber traps nutrients and retards their passage through the GI tract. Exposure of the intestinal mucosa to nutrients induces the release of appetite-regulating peptides which function as hormones or activate neural pathways involved in appetite regulation (Kristensen & Jensen, 2011). (2) Dietary fiber lowers the energy density of a food (Howarth, Saltzman, & Roberts, 2001). Energy density is inversely associated with satiety (Drewnowski, 1998). Thus, by implication, fiber enhances satiety. (3) It requires time and effort to eat the fiber-containing food, increasing mastication. Additionally, fiber stimulates the secretion of saliva and gastric secretions that cause stomach distension, thereby activating satiety signals (Slavin & Green, 2007). (4) Last, colonic fermentation of undigested carbohydrate to short-chain fatty acids has been hypothesized to increase satiety; however, data from human intervention studies do not appear to support a role for intestinal fermentation in appetite regulation (Darzi, Frost, & Robertson, 2011; Hess, Birkett, Thomas, & Slavin, 2011; Peters, Boers, Haddeman, Melnikov, & Qvyjt, 2009).

Highly viscous soluble dietary fiber by increasing the viscosity of GI contents delays gastric emptying which can increase stomach distension (Marciani et al., 2001). Some afferents in the gastric mucosa are mechanoreceptors, while others may stimulate chemical or other signals of satiation (Powley & Phillips, 2004). While gastric satiety is for the most part mechanical in origin, intestinal satiety is nutrient dependent; nevertheless, there exists evidence for a synergy of the two types of stimulation (Maljaars, Peters, & Masclee, 2007; Powley & Phillips, 2004). In the small intestine, transit time is increased and the absorption rate of nutrients is reduced as a result of the increased viscosity of GI contents (Maljaars, Peters, Mela, & Masclee, 2008). A thickening of the unstirred water layer poses an additional barrier to absorption (Johnson & Gee, 1981). Thus, viscous dietary fiber triggers an interaction between neural and hormonal signals that mediate satiety by enhancing the possibility of interaction between nutrients and the cells that release these hormones (Kristensen & Jensen, 2011). Hunger and satiety sensations originate in the central nervous system; however, hormonal secretion from the gut plays a pivotal role in the regulation of food intake (Chaudhri, Field, & Bloom, 2008).

The influence of dietary fiber on body weight is related to its effects not only on satiety and food intake but also on metabolizable energy, which is gross energy minus energy lost in feces, urine, and combustible gases (Lattimer & Haub, 2010). While dietary fiber has been shown to decrease metabolizable energy by decreasing fat digestibility and replacing some simple carbohydrates, the effects of soluble and insoluble fiber on metabolizable energy are the subject of debate. The fat content of the diet and the type of fiber may influence the results (Lattimer & Haub, 2010).

Epidemiologic data suggest that an increase in fiber intake is associated with lower weight gain over the 10- to 12-year period (Liu et al., 2003; Ludwig et al., 1999). An analysis of the data from the Health Professional's Follow-up Study which included 27,082 men followed for 8 years showed that an inverse association existed between dietary fiber intake and weight gain independent of the intake of whole grains. Significant inverse associations were observed between weight gain and fiber from cereals and fruits, but not vegetables. Fiber from fruits displayed the strongest dose–response relationship. Fiber from fruits was associated with reduction in weight gain by 2.51 kg for every 20 g/day increase. However, total dietary fiber was associated with a reduction in long-term weight gain by 5.5 kg, for every 20 g/day increase (Koh-Banerjee et al., 2004). An analysis of data from the EPIC study showed that a 10-g/day intake of dietary fiber was associated with an annual reduction in weight by 39 g and waist circumference by 0.08 cm. Ten grams per day of cereal fiber was associated with an annual weight loss of 77 g/year and a reduction in waist circumference of 0.1 cm/year. Fiber from fruits and vegetables was not associated with an appreciable weight change but was associated with a similar reduction in waist circumference as total fiber (Du et al., 2010).

The effect of dietary fiber on appetite and satiety, energy intake, and body weight as assessed in RCTs has been systematically reviewed (Wanders et al., 2011). Fiber consumption, regardless of the type of fiber, was found to cause an average reduction in appetite by 5% over a 4-hour interval. Further, appetite was determined to reduce by 0.18% per gram of fiber. For more viscous fibers, the reduction increased to 0.41%. Fiber intake also reduced energy intake by 2.6% in studies, wherein fiber supplementation was given over a period of 1 week or more. Irrespective of the type, fiber supplementation was determined to decrease body weight by 0.4% per month (Wanders et al., 2011).

Epidemiologic as well as experimental studies have demonstrated that intake of dietary fiber is inversely associated with satiety, food intake, body weight, and abdominal obesity. While evidence for the effects of fiber from cereals on body weight is supported in epidemiologic studies, the data evaluating the effects of fiber from fruits and vegetables on body weight are inconsistent.

3.2.1 Breads and cereals

Cereals are defined as the fruit of plants that belong to the Gramineae family of grasses and include wheat, rice, barley, corn, rye, oats, millets, sorghum, tef, triticale, canary seed, Job's tears, Fonio, and wild rice. Amaranth, buckwheat, and quinoa function as cereals. However, they are seeds from non-Gramineae families and are referred to as pseudocereals (Harris & Kris-Etherton, 2010). Whole-grain products are derived from cereals and are a good source of dietary fiber. According to the U.S. Food and Drug Administration, to be considered a whole-grain product, the endosperm, germ, and bran components of the grain must be present in the same relative proportion as they are present naturally in the seed. Additionally, whole-grain foods are defined as foods that contain 51% or more whole-grain ingredients (DHHS, 2009).

The 2010 Dietary Guidelines for Americans recommend that at the 2000 kcal level, grain products should comprise six servings of which at least three servings should come from whole grains (USDA-DHHS, 2010). In the United States, total grain servings are typically overconsumed; however, most Americans are not consuming adequate amounts of whole grain (USDA, 2012). An analysis of NHANES data from 1999–2004 indicated that mean whole-grain consumption among adults aged 19–50 and $\geq$51 years was 0.63 and 0.77 servings/day, respectively. Less than 5% of adults in the age group 19–50 years consumed the recommended servings of whole grains (O'Neil, Zanovec, Cho, & Nicklas, 2010). Since the 2005 Dietary Guidelines for Americans, consumers have increased purchases of whole grains, especially cereals, breads, and pasta. Competition among manufacturers leading to more products with whole grains being made available may have triggered the increase in consumption (Mancino, Kuchler, & Leibtag, 2008).

The vast majority of whole grains (56.9%) are consumed at breakfast (Whole-Grain-Council, 2009). Breads (32%) and breakfast cereal products (30%) are the major sources of whole-grain consumption in the United States (Cleveland, Moshfegh, Albertson, & Goldman, 2000), and they have been shown to enhance satiety (Abou Samra, Keersmaekers, Brienza, Mukherjee, & Mace, 2011; Hamedani et al., 2009; Holt, Delargy, Lawton, & Blundell, 1999; Rosen, Ostman, & Bjorck, 2011b; Rosen,

Ostman, Shewry, et al., 2011). However, not all whole-grain breads increased satiety. In a comparison of whole-grain wheat bread and refined-grain wheat bread, subjective satiety and food intake following consumption of the whole-grain bread providing 10.5 g of fiber/day for 3 weeks were not significantly different as compared with refined-grain bread providing 5.8 g of fiber/day (Bodinham, Hitchen, Youngman, Frost, & Robertson, 2011). Fiber components, when added to refined-grain flours used in the production of breads and breakfast cereals, have also been found to have beneficial effects on the regulation of appetite (Lee et al., 2006; Vitaglione et al., 2009). Moreover, enriched and fortified grains provide important nutrients, especially folate (USDA, 2012). Thus, it is vital to encourage consumption of both enriched grains and whole grains in the recommended proportion.

Rye is a good source of soluble and insoluble dietary fiber (Andersson, Fransson, Tietjen, & Aman, 2009). The main fiber components of the cell wall in rye are arabinoxylan, β-glucan, and cellulose. Arabinoxylan is the dominant fiber, and the water-extractable component of arabinoxylan exhibits a high viscosity when dispersed in water (Ragaee, Campbell, Scoles, McLeod, & Tyler, 2001). Arabinoxylan is resistant to the bread-making process and retains its average molecular weight, unlike β-glucan which tends to degrade (Andersson et al., 2009). The molecular weights of the individual fiber types affect their physiologic properties, including viscosity.

Rye flour is usually made from a blend of different rye varieties. Several whole-grain rye breads each made with a different rye variety, including a commercial blend of rye varieties, were compared with bread made from refined-wheat flour. Subjective satiety was significantly higher following consumption of the commercial blend which had the highest insoluble fiber content (10.3 g) as compared with the wheat bread (2.4 g insoluble fiber). However, not all varieties of rye increased satiety (Rosen, Ostman, & Bjorck, 2011b; Rosen, Ostman, Shewry, et al., 2011).

In an assessment of a dose–response relationship, it was found that rye bread (60% rye bran flour, 40% wheat flour) with 5 or 8 g of fiber served as part of isoenergetic breakfasts increased satiety as compared with a wheat bread breakfast. However, there was no significant difference in satiety between the two rye bread breakfasts (Isaksson, Fredriksson, Andersson, Olsson, & Aman, 2009). Varying the structure of rye flour (whole-rye kernels or milled rye kernels) used to make bread did not result in a significantly different effect on satiety (Isaksson et al., 2011).

Rye porridge and rye bread made from different parts of the rye grain (endosperm, whole grain, and bran) were compared with bread made from refined wheat. It was found that the porridge made from whole-grain and bran fractions increased satiety as compared with the bread made from the same parts of the grain; however, all rye products increased satiety when compared with wheat bread (Rosen et al., 2009). The same investigators also compared the effects of similar rye breads on appetite and satiety, with meals made by boiling rye kernels. Besides being more satiating than the breads, the rye kernel meal also resulted in a reduction in food intake at a subsequent meal (Rosen, Ostman, & Bjorck, 2011a).

Whole-grain rye porridge breakfast (followed by whole-grain wheat pasta lunch or refined-wheat pasta lunch) and refined-wheat bread breakfast (followed by refined-wheat pasta lunch) were compared. The meals were matched for macronutrient content. Satiety ratings were significantly higher after the rye porridge breakfast when compared with the refined-wheat bread breakfast. After consuming the refined-wheat pasta lunch meal, subjects who ate the rye porridge breakfast meal continued to have greater sensations of satiety as compared with those who ate the refined-wheat bread breakfast meal (Isaksson, Sundberg, Aman, Fredriksson, & Olsson, 2008). In another study comparing whole-grain rye porridge with an isoenergetic refined-wheat breakfast, satiety was found to be greater after consumption of the rye porridge. Although the effect on satiety was sustained during 3 weeks of regular intake, it was only maintained up to 4 h and energy intake at subsequent meals were not significantly different (Isaksson et al., 2012).

Lupin-kernel flour, derived from the endosperm of lupin seeds contains 40–45% protein and 25–30% fiber with negligible amounts of sugar and starch (Lee et al., 2006). A lupin-kernel fiber-enriched sausage patty has been shown to produce greater effects on satiety than both a conventional patty and an inulin fiber-enriched patty (Archer, Johnson, Devereux, & Baxter, 2004). Partial substitution of lupin-kernel flour for wheat flour used in bread making increases the protein and fiber content of bread. Bread made by a substitution of 40% of wheat flour with lupin-kernel flour was compared with bread made with 100% wheat flour. Served as isoenergetic breakfasts with margarine, and jam, the lupin-kernel fiber bread resulted in greater satiety and lower energy intake at lunch when compared with the wheat bread (Lee et al., 2006).

β-Glucan, which is a soluble fiber found in significant amounts in oats and barley, exhibits a high viscosity at relatively low concentrations (Sadiq Butt, Tahir-Nadeem, Khan, Shabir, & Butt, 2008). The satiating effect of β-glucan has been demonstrated in several studies using β-glucan in doses ranging from 2.2 to 9 g (Beck, Tapsell, et al., 2009; Beck, Tosh, et al., 2009; Lyly et al., 2009; Schroeder, Gallaher, Arndt, & Marquart, 2009; Vitaglione et al., 2009; Vitaglione et al., 2010). However, the results have been inconsistent. Some studies found no effect of β-glucan on satiety (Hlebowicz, Darwiche, Bjorgell, & Almer, 2008; Hlebowicz et al., 2007; Kim, Behall, Vinyard and Conway 2006).

Bread made with 100% wheat flour was compared with bread in which 4.5% of the wheat flour was replaced with 3 g of concentrated extract of barley β-glucan. The bread containing barley β-glucan increased satiety and reduced food intake at a subsequent meal by 19% as compared with the bread made with 100% wheat flour (Vitaglione et al., 2009). In contrast, inclusion of barley β-glucan into breakfast and lunch meals (including barley cereal at breakfast and barley bread at lunch) did not increase satiety as compared with wheat-containing meals (including bran flakes at breakfast and refined-wheat bread at lunch) with similar energy and nutrient contents. Although barley-containing meals were associated with higher energy intake during the remainder of the day, intake was assessed through self-reported food records (Keogh, Lau, Noakes, Bowen, & Clifton, 2007) which are susceptible to misreporting and altered feeding behavior (Stubbs, Johnstone, O'Reilly, & Poppitt, 1998).

Breakfast cereals are often produced from crushed or rolled oats (Sadiq Butt et al., 2008). The content of β-glucan in commercial grade oats in North America varies from 35 to 50 g/kg (Malkki & Virtanen, 2001). When present in cereal-based foods, the physiological response is affected by the amount, solubility, molecular weight, and structure of the β-glucan in the products. These physicochemical properties are in turn affected by the source, processing treatments such as milling, temperature, pH, and shear effects, as well as the interactions with other components in the food matrix (Skendi, Biliaderis, Lazaridou, & Izydorczyk, 2002).

Viscosity is controlled by concentration in solution and molecular weight (Wood, 2007). Oat β-glucan is more soluble in hot water than in water at room temperature, so processing steps that involve moisture and heat will in all likelihood increase the solubility of β-glucan (Tosh et al., 2010). β-Glucan is connected with the cellulose and other noncellulosic polysaccharides in the cell wall and cooking releases it from this matrix (Johansson, Tuomainen, Anttila, Rita, & Virkki, 2006). When used as an ingredient in muffins, the cooking of oats has been shown to increase the percentage of β-glucan solubilized by threefold (Wood, 2004). Thus, food structure and matrix of the product delivering the β-glucan affects its functionality (Skendi et al., 2002).

Breakfast cereals containing oat β-glucan in amounts ranging from 2.2 to 5.7 g and a corn-based breakfast cereal (0 g β-glucan) were compared. The breakfast meals were isoenergetic. Subjective satiety increased with each of the breakfast meals containing oat β-glucan when compared with the corn-based breakfast meal, but there was no difference in the overall satiety responses between the breakfasts containing oat β-glucan in varying amounts. Subsequent food intake decreased only with β-glucan doses in excess of 5 g (Beck, Tosh, et al., 2009). In a separate study, the same investigators examined the effects by varying the dose of oat β-glucan from 2.2 to 5.5 g, delivered through breakfast cereals, on appetite and satiety. They concluded that the optimal dose of β-glucan affecting satiety and other markers of appetite regulation were between 4 and 6 g. Increasing the dose of β-glucan resulted in a greater release of PYY. The hormonal effects were mediated through increased viscosity (Beck, Tapsell, et al., 2009).

In contrast, some studies have shown that oat β-glucan had no effect on satiety. Muesli containing 4 g of oat β-glucan served in yogurt did not increase satiety when compared with an isoenergetic meal consisting of cornflakes served in yogurt (Hlebowicz et al., 2008). Satiety ratings were compared following ingestion of wheat bran flakes (7.5 g fiber), whole-meal oat flakes (4 g fiber: 0.5 g β-glucan), and cornflakes (1.5 g fiber) of equal weight served with milk. Neither bran flakes nor oat flakes resulted in significantly higher satiety when compared with corn flakes (Hlebowicz et al., 2007).

In a study investigating effects of barley β-glucan on satiety, hunger was found to be lower with barley products (9 g β-glucan) when compared with whole-wheat and rice products. The products were served at breakfast as a hot cereal and at mid-morning as a snack mix (Schroeder et al., 2009). In other studies, 1.2 g barley β-glucan in a meal replacement bar (Peters et al., 2009) had no effect on satiety, and a hot cereal containing 2 g of barley β-glucan did not affect short-term satiety in overweight individuals (Kim, Behall, Vinyard, & Conway, 2006).

The insoluble fiber found in breakfast cereals made with whole-grain wheat has also been demonstrated to increase satiety as compared with cornflakes of equal weight (Hamedani et al., 2009) or equal energy content (Samra & Anderson, 2007). The amount of insoluble fiber used in these studies was fairly high, ranging from 26 to 33 g/meal (Hamedani et al., 2009; Samra & Anderson, 2007). In a comparison of isoenergetic breakfasts, increased fullness was observed with a meal high in insoluble fiber (whole-grain wheat bran breakfast cereal: 18.1 g fiber) when compared with a breakfast of bacon and eggs (Holt et al., 1999), higher in fat and comparable in protein.

Whole-grain wheat breakfast cereals that provide fiber in excess of 18 g have been shown to increase satiety, but the evidence is limited. Novel fiber and protein combinations such as that obtained from lupin-kernel flour hold promise, but again the evidence is limited. While the effects of rye on subjective satiety appear to be unequivocal, the results of studies investigating the satiating effects of oats and barley are inconsistent. Additionally, rye porridge appears to be more satiating than rye bread; however, it is not clear if the effects rye exerts on satiety translate into a reduction in food intake.

Epidemiologic as well as intervention trials for the most part evaluated the effects of bread consumption on weight and body composition as part of a food group rather than focusing on bread alone. A comprehensive review (Bautista-Castano & Serra-Majem, 2012) of the epidemiologic data indicated that in a majority of the studies the food group containing bread was not associated with weight status. However, whole-grain bread consumption was found to have beneficial effects on weight status (Cleveland et al., 2000; Greenwood et al., 2000; Koh-Banerjee et al., 2004) and abdominal fat distribution (Halkjaer et al., 2006; Jacobs, Meyer, Kushi, & Folsom, 1998) when compared with bread made with refined grains.

In the Physician's Health Study, a 13-year prospective study, BMI, and weight gain were found to be inversely related to consumption of breakfast cereals regardless of type (whole grain or refined) and independent of other risk factors (Bazzano et al., 2005). In another prospective study, weight gain was inversely associated with the intake of high-fiber, whole-grain foods but positively related to the intake of refined-grain foods (Liu et al., 2003). Using data from NHANES, it was found that eating a breakfast cereal was associated with a significantly lower BMI in adults when compared with other types of breakfast or not eating breakfast (Cho, Dietrich, Brown, Clark, & Block, 2003). Further, consuming a ready-to-eat breakfast cereal (RTEC) was also associated with a macronutrient profile conducive to prevention of obesity in women but not in men (Song, Chun, Obayashi, Cho, & Chung, 2005).

In a RCT, individuals with self-reported night snacking behaviors reduced their postdinner energy intake while consuming a RTEC as an after-dinner snack when compared with a control group consuming their usual snacks (Waller et al., 2004). However, in a study among obese individuals participating in partial meal replacement program, a postdinner RTEC snack did not enhance weight loss when compared with the control group not consuming the RTEC snack (Vander Wal et al., 2006). In another study, while energy intake reduced in the group consuming a RTEC as an evening snack for 6 weeks, there was no significant change in body weight as compared with the control group consuming their usual snacks (Matthews, Hull, Angus, & Johnston, 2012). A RCT tested the effectiveness of a partial meal replacement including breakfast cereal products on weight status. At the end of 4 weeks, greater reductions in BMI, waist, hip, and thigh measurements, as well as percent body fat were seen in subjects consuming the meal replacements containing breakfast cereal products when compared with subjects consuming their normal diet (Wal, McBurney, Cho, & Dhurandhar, 2007). In another study, subjects consuming a RTEC at breakfast and as a meal replacement at lunch or dinner for 2 weeks lost significantly more weight than subjects consuming their usual diet (Mattes, 2002).

Incorporation of oatmeal into a hypocaloric diet for 6 weeks to increase the soluble fiber content (3.9 g) did not result in any significant changes in satiety, body weight, or body composition as compared with a control diet low in soluble fiber (1.9 g) (Saltzman et al., 2001). In an intervention trial to test the effectiveness of weight control measures based on increasing cereal consumption (especially breakfast cereals) or increasing consumption of vegetables, no significant differences in food intake, BMI, or fat mass were observed between the two treatment groups (Rodriguez-Rodriguez et al., 2008). In a 16-week trial to compare the effects of replacing bread, rice, pasta, and breakfast cereals in the usual diet with 100% wheat bread or bread in which 40% of the wheat flour was replaced with lupin-kernel flour, there were no significant effects on body weight or fat mass (Hodgson et al., 2010).

While epidemiologic data suggest that bread consumption does not adversely affect weight status, the paucity of experimental data precludes the establishment of a cause and effect relationship. Observational studies also support whole-grain consumption for improvements in body composition measures. RTECs may have a beneficial role in weight management when used as a partial meal replacement and may reduce energy intake when consumed as a night snack.

3.2.2 Fruits and vegetables

The energy density of a food is primarily a function of its water content; however, fiber does play a lesser role (Drewnowski, 2003). Fat is the most energy-dense nutrient, which provides 9 kcal/g versus 4 kcal/g provided by carbohydrate or protein. Most fruits and vegetables have a high-water content, a low-fat content, and are associated with increased consumption of fiber, making them low in energy density (Rolls, Ello-Martin, & Tohill, 2004). Consumption of fruits and vegetables has been associated with increases in satiety (Gustafsson, Asp, Hagander, & Nyman, 1994, 1995;

Holt et al., 1995) and beneficial effects on weight management (de Oliveira, Sichieri, & Venturim Mozzer, 2008; He et al., 2004; Vioque, Weinbrenner, Castello, Asensio, & Garcia de la Hera, 2008). The energy density and fiber content may have a role to play in the effects of fruit and vegetable consumption on the regulation of food intake and body weight (Alinia, Hels, & Tetens, 2009; Rolls, Ello-Martin, et al., 2004).

The effects of different vegetables on satiety were investigated in a series of experiments (Gustafsson, Asp, Hagander, & Nyman, 1993; Gustafsson et al., 1994, 1995). In the first study (Gustafsson et al., 1993), carrots, peas, Brussels sprouts, or spinach were added to a typical Swedish lunch meal in portions of 96–164 g. The meals were similar in energy and macronutrient content providing 4.4 g of fiber from the added vegetables. No effects of vegetable intake were observed on ratings of satiety as compared with a control lunch. In a subsequent study (Gustafsson et al., 1995) of similar design, 150 and 200 g of spinach providing 4.3 and 7.2 g of fiber, respectively, increased satiety as compared with a control meal without spinach. Further, the satiety ratings positively correlated with the dietary fiber and water content of the meal. When carrots in portions of 100, 200, and 300 g providing 2.9, 5.8, and 8.7 g dietary fiber, respectively, were added to a mixed lunch, satiety increased in a dose–response manner when compared with an isoenergetic meal without carrots (Gustafsson et al., 1994). Thus, it appears that the addition of spinach or carrots to meals in portions of 200 g or more can provoke a satiety response.

In a study comparing the effects of bean purée with potato purée on satiety, it was found that the bean purée delayed the return of hunger and decreased ratings for desire to eat (Leathwood & Pollet, 1988). Another study assessed the effects of a first course comprised of salads (iceberg and romaine lettuce, carrots, cherry tomatoes, celery, and cucumber tossed with Italian dressing and shredded mozzarella and parmesan cheese) in three versions of energy density each served in two different portion sizes. It was found that the portion size of the salad served as a first course was the major factor determining subsequent intake of pasta served as the main course. Energy density had no statistically significant effect on subsequent intake. However, when intake of the entire meal was analyzed, energy intake decreased as the energy density of the salad was decreased, regardless of the portion size. However, portion sizes but not energy density affected satiety ratings with the larger salads eliciting a greater satiety response (Roe, Meengs, & Rolls, 2012). In another study, the covert incorporation of pureed vegetables to lower the energy density (three versions) of main entrees resulted in a decreasing energy intake over the day as the energy density of the entrées was reduced. However, ratings of hunger and fullness did not significantly differ across the conditions (Blatt, Roe, & Rolls, 2011a).

In a study to assess the satiating capacities of isoenergetic portions of 38 different foods categorized into six groups (fruits, bakery products, confectionery, protein-rich foods, carbohydrate-rich foods, and breakfast cereals), fruits were found to elicit the greatest average satiety score (Holt et al., 1995). Although the consumption of dried fruit in the United States is low, it has been associated with improved nutrient intakes, and lower body weight, BMI, and waist circumference (Keast, O'Neil, & Jones, 2011). However, daily consumption of fruits and nut bars (80 g) for 8 weeks did not significantly affect measures of BMI, weight, and waist circumference (Davidi et al., 2011). Nevertheless, a preload of prunes was associated with a greater effect on satiety and a reduction in energy intake at a subsequent meal when compared with an isoenergetic bread and cheese snack (Farajian, Katsagani, & Zampelas, 2010).

Studies have assessed the effects of the physical form of fruits on satiety and energy intake (Bolton, Heaton, & Burroughs, 1981; Flood-Obbagy & Rolls, 2009; Haber, Heaton, Murphy, & Burroughs, 1977). Whole apples were associated with higher satiety ratings than apple purée which in turn was more satiating than apple juice (Haber et al., 1977). Similarly, whole oranges provided greater satiety than orange juice, and whole grapes increased satiety as compared with grape juice (Bolton et al., 1981). A study assessed the effect of a preload of apple juice with added fiber, applesauce, and whole apples matched for weight, energy density, energy, and fiber content on subsequent energy intake. It was found that subjects consumed significantly less energy from the test meal after eating apple segments compared to the applesauce or apple juice, and that the applesauce preload reduced energy intake as compared with the apple juice preload. Eating apple segments also resulted in higher ratings of fullness and lower ratings of hunger compared to other forms of fruits (Flood-Obbagy & Rolls, 2009).

In a review of the epidemiologic data on the relationship between fruit and vegetable intake and body weight, including 16 studies in adults, eight were found to report a significant inverse association between fruit and vegetable intake and weight status which did not vary regardless of the category being fruits and vegetables, fruits only, or vegetables only (Tohill, Seymour, Serdula, Kettel-Khan, & Rolls, 2004). However, in a subsequent review of epidemiologic data, it was concluded that fruit and nonstarchy vegetable consumption was not associated with levels of subsequent weight gain and obesity (Summerbell et al., 2009).

In a prospective investigation involving three separate cohorts, 4-year weight change was found to be inversely related to fruit intake (Mozaffarian, Hao, Rimm, Willett, & Hu, 2011). In the Diet, Obesity, and Genes (DiOGenes) study, including 89,432 individuals from five countries participating in the EPIC study, an inverse association was observed between fruit and vegetable intake and annual weight change (Buijsse et al., 2009). Including participants from 16 centers in addition to those included in the DiOGenes study, the EPIC-PANACEA study found that baseline fruit and vegetable intakes were not associated with weight change after an average of 5 years of follow-up (Vergnaud et al., 2012).

A computer-assisted dieting intervention trial found that although fruit consumption did not increase, fruit intake and body weight were inversely related (Schroder, 2010). In a systematic review, it was determined that among adult experimental studies, increased fruit and vegetable consumption reduced adiposity but the relationship was due to multiple weight-related behaviors. Additionally, in this review, longitudinal studies suggested only a weak relationship between fruit and vegetable consumption and adiposity (Ledoux, Hingle, & Baranowski, 2011). A review of intervention, prospective, observational, and cross-sectional studies on fruit intake and body weight in adults indicated that a majority of the evidence points to an inverse relationship between fruit intake and body weight in the adult population (Alinia et al., 2009).

There appears to be some evidence associating increased fruit consumption with a reduction in adiposity. Fruits and vegetables have a high water, low protein, and low-fat content yet differ greatly in their nutritional profiles, sensory properties, and culinary usage (Dauthy and Food and Agriculture Organization of the United Nations, 1995). The United States Department of Agriculture (USDA) provides separate recommendations for fruits and vegetables (USDA, 2011). Nevertheless, a vast majority of the studies have assessed the combined effects of fruit and vegetable intake on body weight. The data, however, do not support an unequivocal association between fruit and vegetable intake and body weight.

4. FATS AND THE REGULATION OF FOOD INTAKE

The regulation of fat intake involves an integration of physiological events that begins with perception through the nose or mouth of fat-soluble volatile flavor molecules. Food texture defined as the mechanical perception of the oral sensations stimulated by the placement of food in the mouth is then sensed by the oral cavity during chewing and swallowing (Drewnowski, 1997).

However, evidence suggests that the sensing of fat could also be mediated through chemoreception (Drewnowski & Almiron-Roig, 2010). Although the transduction pathways for fatty acid (FA) taste are not clearly understood, numerous receptor systems have been isolated from animal and human tissue (Stewart, Feinle-Bisset, & Keast, 2011).

One mechanism proposed to explain the detection of FAs is through the inhibition of potassium channels (Stewart et al., 2011). It has been shown that long-chain *cis*-polyunsaturated FAs inhibited potassium channels and prolonged stimulus-induced depolarization of rat taste receptor cells. However, saturated, monounsaturated, and *trans*-polyunsaturated fatty acids had no effect on potassium channels (Gilbertson, Fontenot, Liu, Zhang, & Monroe, 1997). Additionally, a FA transporter (CD36) which binds long-chain fatty acids (LCT) in human taste receptors aids in calcium-mediated signaling of taste. The threshold of taste detection is, however, dependent on FA chain length (Mattes, 2009). Using a modified sham feeding technique, in a comparison of high-fat meals containing olive oil, linoleic acid, and oleic acid, it was found that feelings of satiety increased with modified sham feeding of all oils. Thus, the metabolic effects of consumption of high-fat foods may at least in part be modulated by oral exposure (Smeets & Westerterp-Plantenga, 2006).

Lipases present in digestive juices, inside cells, and in endothelial cells aid the chemosensory process. Oxidized FAs or FAs in high concentrations have an unpleasant taste (Stewart et al., 2011). However, in adults, the levels of lingual lipase are low (Drewnowski & Almiron-Roig, 2010). Thus, the levels of fatty acids imputed to stimulate the sensation of taste are low enough to not be sensed as unpleasant, but sufficient to activate taste receptors (Stewart et al., 2011). The ability to taste food containing oxidized fat may be an evolutionary adaptation designed to avoid ingestion of undesirable or toxic compounds (Drewnowski & Almiron-Roig, 2010).

Fat-soluble compounds that contribute to the odor accompanying fats stimulate receptor cells that send signals to brain structures also involved in the processing of emotions and memories. This overlap of neuroanatomical structures offers a biological explanation for feelings of pleasure or disgust that are produced in response to the odorous compounds from fats. Neuroimaging studies have identified the areas of the brain that are activated by fat in the mouth and by viscosity (De Araujo & Rolls, 2004). Further, it appears that the areas of the brain that coordinate the neuronal responses to satiety coincide with the areas of the brain that coordinate neuronal activity related to whether a food tastes pleasant and whether it should be eaten (Rolls, 2004).

Fats are higher in energy density than carbohydrates and proteins, but fats have the added distinction of a characteristic taste and texture that contributes to the palatability of foods (Drewnowski, 1997). One school of thought suggests that palatability reflects an underlying biological need for a nutrient predicted by the sensory properties of the food, while the other relates palatability to reward processes (Yeomans, Blundell, & Leshem, 2004). Distinct neural substrates for homeostatic and hedonic systems have been identified, which implies that the processes of reward can operate free of biological deficits (Blundell & Finlayson, 2004). Palatable foods by influencing appetite sensations can stimulate overconsumption (Yeomans et al., 2004). However, based on this view, unpalatable foods as an appetite-reducing strategy do not present a plausible course of action (Mela, 2006).

Foods that are both energy dense and high in fat are typically the most palatable foods. High palatability is associated with increased food intake, whereas satiety and satiation are associated with a decrease in food intake. Fat by virtue of its palatability stimulates an increase in intake (Drewnowski & Almiron-Roig, 2010). Thus, fat does not satiate but may increase satiety. The combination of taste and smell sensations (using vanilla) has been shown to enhance satiety following consumption of a high-fat meal (Warwick, Hall, Pappas, & Schiffman, 1993). However, the role of palatability in fat-induced overeating is unclear. While the initial food selection may be based on orosensory qualities, postingestive nutritional factors may determine how much energy is consumed (Sclafani, 2004). Children given repeated exposures to distinctly flavored high-fat and low-fat yogurt drinks matched for orosensory characteristics increased their preference for the high-fat flavor (Johnson, McPhee, & Birch, 1991).

Fats have been shown to reduce hunger when present in the GI tract by eliciting satiety signals (Little & Feinle-Bisset, 2011). Fat in the duodenum stimulates the release of cholecystokinin directly and other GI peptides such as PYY and GLP-1 by an indirect neurohumoral pathway to affect satiety (Maljaars et al., 2007). Exposure of the ileum to fat stimulates an even larger satiety response than exposure to the duodenum (Maljaars et al., 2008). Fat reaching the ileum stimulates the ileal brake, a distal to proximal feedback mechanism that slows gastric emptying and delays the transit of food through the GI tract. Nutrients in the small intestine influence satiety and food intake by activation of neural afferents or by inducing the release of gut hormones involved in appetite regulation (Maljaars et al., 2007; Van Citters & Lin, 1999).

Bariatric surgery is arguably the most effective weight-loss treatment for the morbidly obese (Karra, Chandarana, & Batterham, 2009). The

Roux-en-Y gastric bypass surgery results in a speedy delivery of partially digested nutrients to the distal parts of the GI tract. Meal-stimulated increases in PYY, and GLP-1, gut hormones with anorectic effects, implicated in the ileal brake activation, have been observed in subjects who have undergone the Roux-en-Y gastric bypass (Field, Chaudhri, & Bloom, 2010; Karra et al., 2009). Thus, the reduction in body weight and the physiological responses observed following bariatric surgery provide evidence that a sustained appetite-reducing effect is possible through a recurring activation of the ileal brake (Maljaars et al., 2008).

Infusion of triglycerides into the ileum has been shown to alter duodenal motility and delay gastric emptying (Fone, Horowitz, Read, Dent, & Maddox, 1990). Ileal fat infusion has also been shown to cause a dose-dependent delay in gastric emptying and has been related to increased plasma concentrations of PYY (Pironi et al., 1993; Read et al., 1984). Following an ileal infusion of corn oil, feelings of satiety increased and *ad libitum* food intake decreased at a meal 30 min after the start of the infusion. Although in this study, the rate of infusion of fat can be compared to what one may find in normal subjects after eating a heavy meal (Welch, Saunders, & Read, 1985; Welch, Sepple, & Read, 1988), even a low physiological dose of fat (6 g) into the ileum elicited a significant reduction in hunger and food intake when compared with an oral ingestion of the same amount of fat (Maljaars et al., 2011).

In a pooled analysis of studies investigating the effects of fat on gastric emptying and GI hormone release, it was determined that the magnitude of stimulation of pyloric pressures and release of cholecystokinin, a hormone with anorexigenic effects, are independent predictors of subsequent energy intake (Seimon et al., 2010). A high-fat breakfast meal has been shown to delay gastric emptying at lunch as compared with low-fat meals matched for energy or mass of the high-fat meal; however, the high-fat breakfast meal resulted in increased food intake 7 h later (Clegg & Shafat, 2010).

The regulation of GI motor function, gut hormone release, and satiety by fat is affected by its physicochemical properties. These effects are more pronounced with LCT ($\geq$12 carbons) than shorter chain fatty acids (Feltrin et al., 2004; French et al., 2000; Little & Feinle-Bisset, 2011). Food intake was reduced by over 200 kcal following a duodenal infusion of long-chain fat emulsions (180 kcal) when compared with a saline infusion (French et al., 2000). Duodenal infusion of 12-carbon fatty acids reduced appetite and energy intake at a subsequent meal when compared with 10-carbon fatty acids. The effects on gastroduodenal motility that were observed are typically associated with delayed gastric emptying (Feltrin et al., 2004). It has been suggested that accelerated gastric emptying decreases gastric distension, thereby promoting hunger (Little, Horowitz, & Feinle-Bisset, 2007). Thus, hunger and gastric emptying are closely related. For fat to affect gastric emptying, however, digestion of fats and consequent release of free fatty acids appear to be crucial (Little et al., 2007).

Medium chain triglycerides (MCT) (6–12 carbons) have been shown to influence satiety through increased energy expenditure. Unlike LCT, MCT are directly absorbed into portal circulation and are more rapidly metabolized. The faster rate of oxidation increases thermogenesis (St-Onge & Jones, 2002). Three isoenergetic breakfasts matched for fat content, but differing in fatty acid chain length was compared with respect to their effects on satiety. LCT from beef tallow, MCT from coconut oil, and short-chain triglycerides from dairy fat were added to savory muffins. There was no significant difference in satiety ratings following consumption of the three breakfasts differing in the type of lipid when measured over 6 h (Poppitt et al., 2010).

In a comparison of meals differing in the degree of saturation of the fat content, no significant differences in satiety were observed. The meals were high in polyunsaturated fatty acids from walnuts, or monounsaturated fatty acids from olive oil, or saturated fatty acids from dairy fat (Casas-Agustench et al., 2009). The role played by the degree of saturation in modulating the effects of fat on the GI tract has yet to be resolved (Maljaars, Romeyn, Haddeman, Peters, & Masclee, 2009; Strik et al., 2010).

Pinnothin™ is a natural oil pressed from Korean pine nuts and contains linoleic acid (C18:2), pinolenic acid (C18:3), and oleic acid (C18:1). Consumption of Pinnothin™ triacylglycerols and free fatty acids have been shown to produce an increase in cholecystokinin and GLP-1 in postmenopausal overweight women when compared with olive oil. However, appetite ratings did not significantly differ (Pasman et al., 2008). In overweight women, although consumption of Pinnothin™ free fatty acids reduced food intake by 7% at a subsequent meal, appetite ratings were not significantly different after consumption of Pinnothin™ triacylglycerols or free fatty acids compared to olive oil (Hughes et al., 2008). In both the studies (Hughes et al., 2008; Pasman et al., 2008), participants consumed Pinnothin™ in a capsule form. Added to yogurt, Pinnothin™ triacylglycerol consumption did not result in appetite sensations and energy intake that were significantly different when compared with milk fat (Verhoef & Westerterp, 2011).

Delaying lipid digestion is an important factor in stimulating the ileal brake. The digestion of fat can be slowed down by manipulating the oil emulsion interfacial composition using galactolipids. It has been shown that galactolipids reduce the rate and extent of lipolysis by sterically hindering the penetration of pancreatic colipase and lipase or preventing the formation of a colipase–lipase complex at the oil–water interface in the duodenum (Chu et al., 2009). Olibra™ is a fat emulsion comprising fractionated palm and oat oil in the proportion of 95:5. The palm oil is emulsified by hydrophilic galactolipids derived from oat oil (Knutson et al., 2010). One study using a method of delivering Olibra™ directly into the GI tract demonstrated a delay in GI transit (Knutson et al., 2010). Although oral administration in another study showed a 45-min delay in orocecal transit time (Haenni, Sundberg, Yazdanpandah, Viberg, & Olsson, 2009), the computation of orocecal transit time has been questioned (Peters, Beglinger, Mela, & Schuring, 2010). However, when ingested orally, the GI responses manifested by an intragastric administration may differ. Unless the emulsion is resistant to digestion in the dynamic environment of the GI tract, an increase in satiety and a reduction in food intake are unlikely to occur.

Early studies (Burns, Livingstone, Welch, Dunne, & Rowland, 2002; Burns et al., 2000; Burns et al., 2001), all crossover designs, reported a reduction in energy, macronutrient, and total weight of food intake following consumption of yogurt containing the Olibra™ emulsion. Subsequent studies failed to show a reduction in energy intake (Chan et al., 2012; Diepvens, Steijns, Zuurendonk, & Westerterp-Plantenga, 2008; Logan et al., 2006; Rebello, Martin, Johnson, O'Neil, & Greenway, 2012). Olibra™ has been shown to positively impact body composition and weight maintenance after weight loss (Diepvens, Soenen, Steijns, Arnold, & Westerterp-Plantenga, 2007), but in another study although body fat mass decreased by 0.9%, there was no change in body weight at the end of 12 weeks (Olsson, Sundberg, Viberg, & Haenni, 2011). In these studies by Diepvens et al. and Olsson et al., the calorie restriction imposed during the weight-loss period may have had a role to play in the beneficial effects. In a recent review, it was concluded that Olibra had no efficacy as a satiety-enhancing weight-loss strategy (Rebello et al., 2012).

In humans, exposure to a high-fat or high-energy diet has been shown to decrease sensitivity to the GI mechanisms that regulate appetite (Clegg et al., 2011; Little & Feinle-Bisset, 2011; Little et al., 2007). It has been suggested that dietary restriction may cause a reversal of these effects resulting in enhanced nutrient sensing and appetite suppression (Little & Feinle-Bisset, 2011). A modification of appetite perceptions with an increase in hunger and a decrease in fullness has been observed following a high-fat diet

(58% of energy intake) for 2 weeks. Further, a significant increase in energy intake of about 160 kcal/day was observed for the following 2-week period (French, Murray, Rumsey, Fadzlin, & Read, 1995). Placing subjects on a high-fat diet derived from sunflower oil for only 3 days resulted in an acceleration of gastric emptying (Clegg et al., 2011). However, the acceleration in GI transit and reduction in satiety following a high-fat diet that occurred over a 1-week period returned to prediet levels by the end of 4 weeks (Clegg et al., 2011).

Fat is the most energy-dense macronutrient, contributing to pleasantness and thereby perceived palatability of foods, which may induce over-consumption. Fats bestow on foods a wide range of taste and texture properties, making it difficult to determine which particular oral sensations contribute to the perception of fat content. Fat perception is also influenced by physical form and other taste sensations, such as sweetness. Important textural properties include viscosity and lubricity (Drewnowski & Almiron-Roig, 2010). By adding hydrocolloid thickeners or other components that influence viscosity, it is possible to create an illusion of fat content (Drewnowski & Almiron-Roig, 2010). While fat may not reduce meal termination (satiation), fat in the GI tract generates satiety (meal initiation). The physicochemical properties of fat influence the postingestive effects of fat on satiety, but these properties can be manipulated. Thus, the paradoxical effects of fats on energy density and satiety notwithstanding, they could be manipulated to produce a desired directional change in feeding behavior.

5. TEAS, CAFFEINE, AND PUNGENT FOODS

Certain food components do not provide energy but have been shown to increase energy expenditure. Tea is made from the leaves of the *Camellia sinensis* L. species of the Theaceae family. Oolong tea is partially fermented and oxidized, while green tea is not fermented or oxidized. Both oolong and green tea contain several polyphenolic components such as epicatechin, epicatechin gallate, epigallocatechin, epigallocatechin gallate (EGCG), and caffeine. Of these polyphenols, EGCG is the most abundant and is highly active pharmacologically (Hursel & Westerterp-Plantenga, 2010; Kovacs & Mela, 2006).

Catechins inhibit catechol O-methyltransferase an enzyme that degrades norepinephrine, and caffeine inhibits phosphodiesterase, an enzyme that degrades c-AMP. A reduction in degradation causes an increase in the levels of norepinephrine and c-AMP. Norephinephrine controls biochemical mechanisms that either result in an increased use of ATP or an increased rate of mitochondrial oxidation with inefficient coupling of ATP synthesis, leading to increased heat production (Hursel & Westerterp-Plantenga, 2010; Westerterp-Plantenga, Diepvens, Joosen, Berube-Parent, & Tremblay, 2006). Thus, the thermogenic effects of caffeine and tea catechins are related to prolonged or increased stimulatory effects of norepinephrine and c-AMP on energy and lipid metabolism (Kovacs & Mela, 2006).

The results of a meta-analysis of studies investigating the effects of green tea on weight loss and weight maintenance suggest that an EGCG–caffeine mixture has a beneficial effect on weight loss and weight maintenance after a period of energy restriction. Subjects in the treatment groups lost an average of 1.31 kg or gained 1.31 kg less weight than subjects in the control groups over a 12-week period. However, habitually low-caffeine consumers reacted with greater sensitivity than habitually high consumers. Ethnicity appeared to be a moderator of the thermogenic effect as Asian subjects lost more weight than Caucasians. There was no dose–response relationship between intake of catechins and body weight (Hursel, Viechtbauer, & Westerterp-Plantenga, 2009).

Caffeine belongs to a class of compounds called methylxanthines and is present in coffee, tea, cocoa, chocolate, and some cola drinks (Westerterp-Plantenga et al., 2006). The effect of caffeine intake on energy expenditure has been demonstrated in several short-term studies (Acheson, Zahorska-Markiewicz, Pittet, Anantharaman, & Jequier, 1980; Acheson et al., 2004; Arciero, Gardner, Calles-Escandon, Benowitz, & Poehlman, 1995; Astrup et al., 1990; Bracco, Ferrarra, Arnaud, Jequier, & Schutz, 1995; Dulloo, Geissler, Horton, Collins, & Miller, 1989; Hollands, Arch, & Cawthorne, 1981). However, a RCT found that there was no effect on body weight over a 16-week period (Pasman, Westerterp-Plantenga, & Saris, 1997) although epidemiologic data from a 12-year prospective study supported an inverse association between caffeine intake and long-term weight gain (Lopez-Garcia et al., 2006). It appears that although caffeine intake may result in increasing energy expenditure in the short term, the evidence to support its effects on weight loss is lacking. The available evidence supports a role for green tea in weight loss; however, a meta-analysis determined that the magnitude of the effect may lack clinical relevance (Phung et al., 2010).

Capsaicin is the major pungent ingredient in red hot pepper. Capsaicin has been reported to increase thermogenesis by enhancing catecholamine secretion and inducing β-adrenergic stimulation (Yoshioka et al., 1995). Capsaicin is perceived as pungent because it activates the transient receptor potential vanilloid receptor 1 (TRPV1) found in neurons on the tongue.

The activation of the TRPV1 receptor stimulates the release of catechol-amines which leads to an increase in energy expenditure by stimulation of the sympathetic nervous system and the upregulation of uncoupling proteins (Hursel & Westerterp-Plantenga, 2010).

Studies have assessed the effects of capsaicin on energy metabolism in humans and demonstrated an increase in energy expenditure (Yoshioka et al., 1995), diet-induced thermogenesis, and fat oxidation (Yoshioka, St-Pierre, Suzuki, & Tremblay, 1998). In a comparison of high-fat and high-carbohydrate breakfast meals with and without red pepper, it was found that the red pepper-containing meals reduced appetite before lunch. Differences in diet composition at the breakfast meal did not affect energy and macronutrient intake at lunch, but protein and fat intake at lunch was reduced with intake of red pepper-containing meals consumed at breakfast (Yoshioka et al., 1999). The effects were more pronounced in the high fat as opposed to the high-carbohydrate diet (Yoshioka et al., 1998; Yoshioka et al., 1999). The addition of red pepper to an appetizer at lunch time resulted in reduced intake of carbohydrate and energy during the rest of the lunch and a snack served several hours later (Yoshioka et al., 1999). The effect of red pepper on reducing energy intake was found to be greater when administered in tomato juice than when administered in capsule form (Westerterp-Plantenga, Smeets, & Lejeune, 2005). In another study, although a capsaicin containing meal resulted in an increase in GLP-1, there were no effects on appetite and energy expenditure as compared with a control meal without capsaicin (Smeets & Westerterp-Plantenga, 2009).

The data on the long-term consumption of capsaicin are scarce. Capsaicin supplementation for 3 months after a modest weight loss had no effect on weight maintenance as compared with a control group that was not supplemented. In this study, compliance with the diet appeared to pose a problem (Lejeune, Kovacs, & Westerterp-Plantenga, 2003). Capsinoids are nonpungent capsaicin-related substances found in the CH–19 sweet pepper and have been investigated for their thermogenic effects (Snitker et al., 2009). Capsinoid supplementation for 12 weeks was well tolerated but did not affect energy expenditure or body weight, although a significant increase in fat oxidation and reduction in abdominal adiposity was observed (Snitker et al., 2009).

There appears to be some evidence to support an increase in satiety and a reduction in food intake following consumption of foods containing capsaicins but the data are inconsistent. Long-term studies investigating the effects of capsaicin consumption on body weight are lacking which may perhaps be due to the difficulty in adhering to a diet containing pungent foods. The CH-19 sweet pepper may promote greater compliance, but its effects on satiety, food intake, and body weight need further substantiation through controlled studies.

6. ENERGY DENSITY

Energy density is defined as the amount of energy per unit weight of a food or beverage (most commonly expressed as kilocalories per gram or kilojoules per gram). The amount of water present in a food is a major influencer of energy density because water adds weight without adding calories. Macronutrient composition also influences the energy density of a food with fat providing 9 kcal/g compared to 4 kcal/g for carbohydrates and protein. Dietary fiber adds weight while contributing minimal energy. Thus, food that is high in water and/or fiber is often low in energy density.

Several epidemiologic studies have found a positive association between the energy density of the diet and measures such as weight gain, BMI, and waist circumference (Perez-Escamilla et al., 2012). While the effects of energy density on satiety have been mixed, numerous short-term studies have found that increasing the energy density of a test food or meal decreases energy intake at subsequent meals. Some studies have shown that adding a low-energy density preload, such as salad or soup, before a meal decreases the amount of food consumed at the meal (Flood & Rolls, 2007; Rolls, Bell, & Waugh, 2000; Rolls, Roe, & Meengs, 2004; Rolls et al., 1998). Flood and Rolls found that when subjects consumed vegetable soup prior to their lunch, meal-time energy intake decreased by 20% or 134 ± 25 kcal (Flood and Rolls, 2007). Consuming salad before or with a meal resulted in an 11% (57 ± 19 kcal) decrease in meal-time energy intake (Roe et al., 2012). These studies demonstrate that adding low-energy density foods to a meal can reduce the amount of calories consumed at a single meal.

One strategy to lower the energy density of foods is to increase their water content. Work from Barbara Rolls' lab has shown that incorporating water into food decreases meal-time energy consumption (Rolls, Bell, & Thorwart, 1999). Adding water to a chicken rice casserole so that it became a chicken rice soup resulted in a 16% reduction in energy intake at lunch. Subjects did not compensate for their reduced lunch-time food intake at dinner. However, water served as a beverage with a meal did not alter energy intake suggesting that water as a beverage is perceived differently than water in a food such as soup.

Another strategy for reducing the energy density of foods is the incorporation of puréed vegetables into recipes. Blatt et al. covertly substituted puréed carrots, squash, and cauliflower into various recipes and tested subjects' energy intake compared to the normal recipes. When the energy density of meals was lowered by 15%, daily energy intake decreased by 202 ± 60 kcal. When the energy density of meals was lowered by 25%, daily energy intake decreased by 357 ± 47 kcal (Blatt et al., 2011a). While ingredient substitution may be challenging to implement for the home cook, it has been shown to be effective for decreasing short-term energy intake and increasing the amount of vegetables eaten. Various other studies have found that decreasing the energy density of a meal leads to decreased energy intake at the meal itself and subsequent food intake later in the day (Bell, Castellanos, Pelkman, Thorwart, & Rolls, 1998; Bell & Rolls, 2001; Chang, Hong, Suh, & Jung, 2010; Cheskin et al., 2008; Latner, Rosewall, & Chisholm, 2008; Rolls, Bell, Castellanos, et al., 1999; Rolls, Roe, & Meengs, 2006). Collectively these studies show that lowering the energy density of meals can successfully reduce short-term energy intake, which suggests that maintaining a low-energy density diet may result in weight loss. While the evidence from short-term food intake studies is very strong, longer-term weight-loss studies have produced mixed results.

A small number of RCTs have been performed to examine the role of energy density in weight loss. Two studies manipulated the diet of subjects by having them add snacks of varying energy density to their diets. De Oliveira et al. had overweight or obese subjects ($n = 49$) add three apples, pears, or oat cookies to their usual diet and measured weight change over 10 weeks (de Oliveira et al., 2008). Both the apple and pear groups lost weight (-0.93 and -0.84, respectively), while the oat cookie group gained a small amount of weight ($+0.21$ kg). Viskaal-van Dongen et al. had normal-weight participants add low or high-energy density snacks to their usual diets but found no differences in weight change between the two groups after 8 weeks (Viskaal-van Dongen, Kok, & de Graaf, 2010).

Alternatively, other trials have provided counseling on increasing the energy density of participants' diets but provided no study foods. A short 4-week study found no differences between a calorie-restricted high-energy density diet and a calorie-restricted low-energy density diet (Song, Bae, & Lee, 2010). The low-energy density group reported less hunger but weight loss was not different between the groups. A 12 week study found that subjects on a low-energy density diet lost an average of 9.3 kg similar to results achieved by subjects consuming an energy-restricted low-fat diet (-7.7 kg). Combining these two sets of dietary advice did not result in enhanced weight loss (Raynor, Looney, Steeves, Spence, & Gorin, 2012). Similarly, a 6-month weight-loss trial for obese women produced an average weight change of −6.4 kg with a reduced-fat diet and −7.9 kg with a reduced-fat diet plus increased fruits and vegetable consumption (Ello-Martin, Roe, Ledikwe, Beach, & Rolls, 2007). Though the 1.5 kg difference is modest, these values were statistically significant ($P=0.019$). The longest study which measured weight change with reduced dietary energy density is a 4-year trial with female breast cancer survivors. Groups received counseling on increasing fruit and vegetable consumption or general dietary guideline materials. After 4 years, there were no differences in weight between the two groups (Saquib et al., 2008). Overall, the evidence suggests that reducing dietary energy density may be an effective tool for promoting weight loss. However, there currently is no evidence to suggest that it is superior to other weight-loss strategies.

Choosing foods with lower energy density may be one way to promote reduced energy intake and enhance weight loss, but there are some exceptions to choosing solely based on energy density. Sugar sweetened beverages such as soda generally have a low-energy density value, but they have been linked to the promotion of excess energy intake and weight gain (Malik, Popkin, Bray, Despres, & Hu, 2010). Indeed, several studies have found that beverages are only weakly satiating compared to solid food, and it may be that calories consumed from beverages are not sensed by the body in the same way as calories from foods (Mattes, 2006; Mattes & Campbell, 2009). The other exception appears to be nuts, which were discussed earlier in this review. Nuts have very high-energy density values but do not appear to cause excess weight gain when consumed regularly. The energy densities of some common foods are presented in Table 3.1.

Table 3.1 Energy density of selected foods based on the United States Department of Agriculture National Nutrient Database for Standard Reference, Release 24

Food	Energy density (kcal/g)
Butter	7.17
Walnuts	6.54
Almonds	5.95
Peanuts	5.85
Pistachio nuts	5.67
Potato chips	5.42

Table 3.1 Energy density of selected foods based on the United States Department of Agriculture National Nutrient Database for Standard Reference, Release 24—cont'd

Food	Energy density (kcal/g)
Chocolate chip cookie	4.54
Cheddar cheese	4.03
Prunes	3.39
Pork sausage	3.39
Cheesecake	3.21
Pizza	2.76
Ground beef	2.70
Rye bread	2.58
Bagel	2.57
Whole-wheat bread	2.47
Chicken breast	1.65
Chickpeas	1.64
Eggs	1.55
Navy beans	1.40
Tuna	1.28
Lentils	1.16
Low-fat yogurt	1.02
Banana	0.89
Tofu	0.70
Apple	0.52
Orange juice	0.49
1% milk	0.42
Carrots	0.41
Butternut squash	0.40
Grapefruit	0.32
Cauliflower	0.25
Spinach	0.23
Lettuce	0.15

7. MEAL PLANS

The effects of select foods on satiety, food intake, and body weight using evidence from RCTs are presented in Table 3.2. Meal plans that provide 1200, 1600, or 2000 kcal were developed (Tables 3.3–3.5). These meal plans meet 100–113% of the USDA recommendations in the dairy, fruit, vegetable, and grain food groups and provide about 22–24% of energy from protein. Despite the relatively high-protein contents, the meal plans do not exceed one egg/day and 6 ounces/day in servings from meat and fish. Dairy products are either low fat or fat free in keeping with the recommendations

Table 3.2 The effects of individual foods on satiety, food intake, and body weight using evidence from randomized controlled trials

Food	Satiety	Food intake	Body weight
Protein	+	+/−	+/−
Dairy products	ND	ND	+/−
Milk	+/−	+/−	+ (with energy restriction)
Yogurt	+	−	+
Cheese	ND	ND	ND
Meat and meat products	ND	ND	−
Beef/pork/chicken	−	−	−
Fish	+	+	+
Eggs	+	+/−	+ (with energy restriction)
Pulses	+	ND	+/−
Chickpeas	−	−	ND
Lentils	+/−	+/−	ND
Navy beans	−	−	ND
Yellow peas	+/−	+/−	ND
Soybean	+/−	−	−
Walnuts	+/−	−	+

Table 3.2 The effects of individual foods on satiety, food intake, and body weight using evidence from randomized controlled trials—cont'd

Food	Satiety	Food intake	Body weight
Almonds	+	+	+/−
Peanuts	ND	ND	+
Pistachios	ND	ND	+
Carbohydrates			
Breads and cereals	ND	ND	+/−
Whole-wheat bread	−	−	ND
Rye bread	+	ND	ND
Lupin bread	+	+	ND
Barley bread	+/−	+/−	ND
Rye porridge	+	+/−	ND
Oat breakfast cereal	+/−	−	ND
Barley breakfast cereal	+/−	−	ND
Whole-wheat breakfast cereal	+	ND	ND
Ready-to-eat-cereal	ND	+	+/−
Fruits and vegetables	ND	ND	−
Fruits	+	ND	+
Apple/pear/grapefruit	ND	ND	+
Dried fruit/prunes	+	+	+
Vegetables	+	ND	ND
Spinach ≥ 200 g	+	ND	ND
Carrots ≥ 200 g	+	ND	ND
Salad (raw vegetables)	ND	+	ND
Teas, caffeine, and pungent foods			
Green tea	ND	ND	+
Oolong tea	ND	ND	+
Caffeine	ND	ND	−
Capsaicin	+/−	+/−	−

+, Beneficial effect; − no effect; +/− inconsistent effect; ND, not determined.

Table 3.3 Meal plan that provides approximately 1200 kcal/day

Meal	1200 kcal	Weight (g)
Breakfast	1 cup cooked rye porridge or oatmeal (a, b)	234
	¼ cup dried fruit (a, b)	34
	1 cup fat-free milk (a, b, c)	245
Snack	6 ounces low-fat yogurt (a, c)	183.75
	1 cup green tea (c)	
Lunch	1 egg omelet (a, b, c)	61
	Sautéed green beans ¾ cup cooked green beans (a)	93.75
	1 teaspoon olive oil	4.5
	¼ cup sliced almonds (a, b, c)	26.25
	1 slice bread (rye, lupin, or barley) (a, b)	32
Snack	½ cup fruit (a, c)	77.63
	1 cup fat-free milk (a, b, c)	245
Dinner	1½ cups chicken and rice soup (d)	361.5
	1 slice bread (rye, lupin, or barley) (a, b)	32
	½ cup steamed carrots sliced (a)	78

The meal plan meets 100–112% of the USDA recommendations in the dairy, fruit, vegetable, and grain food groups and provides 24% of energy from protein. Energy density of the diet is 0.70 kcal/g. Analyzed using the USDA National Nutrient Database for Standard Reference, Release 24.
a, Satiety; b, food intake; c, body weight: evidence provided in randomized controlled trials. d, Energy density studies: Rolls et al. (1998, 2000b), Rolls, Roe, and Meengs (2004), Flood and Rolls (2007). Green tea is not included in the energy density calculation.

Table 3.4 Meal plan that provides approximately 1600 kcal/day

Meal	1600 kcal	Weight (g)
Breakfast	1 egg (a, b, c)	61
	2 slices bread (lupin, rye, or barley) (a, b)	64
	2 teaspoons light *trans*-fat-free margarine	9.6
	1 slice low-fat cheese	28.35
	1 cup fruit (a, c)	155.27
Snack	6 ounces low-fat yogurt (a, c)	183.75
	¼ cup dried prunes (a, b, c)	43.5
	1 cup green tea (c)	

Table 3.4 Meal plan that provides approximately 1600 kcal/day—cont'd

Meal	1600 kcal	Weight (g)
Lunch	Grilled chicken salad (d)	
	2 cups lettuce/tomatoes/celery/cucumber (b)	237
	2 tablespoons low-fat salad dressing	32
	2 ounces grilled chicken breast	56.7
	¼ cup sliced almonds (a, b, c)	26.25
	1 slice bread (lupin, rye, or barley) (a, b)	32
Snack	1 cup whole-wheat cereal (a)	60
	1½ cups fat-free milk (a, b, c)	367.5
Dinner	4 ounces fish (steamed or grilled) (a, b, c)	113.4
	1 cup sautéed spinach (a)	180
	1 teaspoon olive oil	4.5
	1 slice bread (lupin, rye, or barley) (a, b)	32

The meal plan meets 100–109% of the USDA recommendations in the dairy, fruit, vegetable, and grain food groups and provides 24% of energy from protein. Energy density of the diet is 0.95 kcal/g. Analyzed using the USDA National Nutrient Database for Standard Reference, Release 24.
a, Satiety; b, food intake; c, body weight: evidence provided in randomized controlled trials. d, Energy density studies: Rolls et al. (1998, 2000b), Rolls, Roe, and Meengs (2004, 2007). Green tea is not included in the energy density calculation.

of the 2010 Dietary Guidelines for Americans. The fiber content ranges from 20 g/day in the 1200 kcal diet to 42 g/day in the 2000 kcal diet. Nonnutritive sweeteners may be added to increase the palatability of foods, if desired.

The most distinctive aspect of these meal plans is the evidence-based consideration given to satiating properties of the foods included, and the overall energy density of the diet. We would like to term these meals as a "high satiety" (HS) plan. To illustrate the basis for choosing various foods to include in the HS meals, we have used a scoring system. The letters a–d following the foods indicate the underlying available research evidence about that food. For instance, the letter a denotes that the food was reported to induce a subjective feeling of satiety, the letter b indicates evidence for a reduction in food intake, c refers to evidence for a role in weight loss, and the letter d indicates evidence for using that food to increase satiety by lowering

Table 3.5 Meal plan that provides approximately 2000 kcal/day

Meal	2000 kcal	Weight(g)
Breakfast	1 cup rye porridge or oatmeal (a, b)	234
	¼ cup sliced almonds (a, b, c)	28.35
	1 egg (a, b, c)	61
	1 slice bread (lupin, rye, or barley) (a, b)	32
	1 teaspoon light *trans*-fat-free margarine	4.8
	1 cup fat-free milk (a, b, c)	245
Snack	½ cup bean dip (a, c)	131
	6 baked tortilla chips	8.4
	1 cup green tea (c)	
Lunch	Roast beef sandwich	
	2 ounces roast beef	56.7
	½ cup lettuce + 2 tomato slices (b)	68
	1 tablespoon light mayonnaise	15.6
	2 slices bread (lupin, rye, or barley) (a, b)	64
	1 cup lentil soup (a, b, d)	248
	½ cup grapefruit (c)	115
Snack	1 medium pear (c)	178
	1½ cups fat-free milk (a, b, c)	367.5
Dinner	4 ounces fish (baked or broiled) (a, b, c)	113.4
	Spinach salad (d)	
	2 cups raw spinach (a, b)	60
	¼ cup dried cranberries (a, b)	27.5
	2 tablespoons low-fat salad dressing	32
	¼ cup low-fat cheese crumbled	37.5
	1 slice bread (lupin, rye, or barley) (a, b)	32
	1 teaspoon light *trans*-fat-free margarine	4.8

The meal plan meets 100–113% of the USDA recommendations in the dairy, fruit, vegetable, and grain food groups and provides 22% of energy from protein. Energy density of the diet is 0.92 kcal/g. Analyzed using the USDA National Nutrient Database for Standard Reference, Release 24.
a, Satiety; b, food intake; c, body weight: evidence provided in randomized controlled trials. d, Energy density studies: Rolls et al. (1998, 2000b), Rolls, Roe, and Meengs, (2004, 2007).

the energy density. Thus, at a glance, these HS meal plans convey the scientific basis for the foods included. The value of the HS-meal plans is apparent, particularly when compared with the typical American meal plan (Table 3.6). Based on the evidence, almost every food item included in the HS plans has satiating properties compared to only two items (raw vegetables and whole milk) in the typical American plan. The energy density of the HS meals ranges from 0.70 to 0.95 kcal/g compared to 1.54 kcal/g for the typical American diet.

Table 3.6 Sample menu of the typical American diet (2100 kcal/day) based on the control diet used in the DASH trial[a]

Meal	2100 kcal	Weight (g)
Breakfast	Apple juice unsweetened	126
	Blueberry muffin	50
	Butter without salt	10
	Jelly	14
	Whole milk	120
Lunch	Turkey breast meat	100
	Lettuce, iceberg (raw)	20
	Mayonnaise salad dressing	20
	Bread (white)	55
	Yellow cake	50
	Chocolate frosting	25
Dinner	Pork stir fry	227
	Olive oil	6
	Rice (cooked, white)	200
	Bread (French)	40
	Butter	20
	Gelatin dessert	135
	Dessert topping (nondairy)	20
Snack	Applesauce (canned)	110
	Saltine crackers	20

Energy density of the diet is 1.54 kcal/g.
[a]Appel et al. (1997).

We would like to emphasize a few points. Most of the foods included in these HS-meal plans have been shown to increase satiety when studied individually. The evidence relating to the satiety-enhancing effect is neither unequivocal in each case nor is there evidence to demonstrate that when consumed collectively, the satiating foods will have a synergistic or additive effect on satiety. We have postulated this logical extension of the available information. Perhaps, future research may test these concepts.

In summary, based on the available data and information, we have created HS-meal plans. Compared to a typical American diet, these meal plans are considerably lower in energy density and are probably more satiating. A diet that exploits the satiating properties of multiple foods may help increase long-term dietary compliance, and consequentially enhance weight loss.

REFERENCES

Abargouei, A. S., Janghorbani, M., Salehi-Marzijarani, M., & Esmaillzadeh, A. (2012). Effect of dairy consumption on weight and body composition in adults: A systematic review and meta-analysis of randomized controlled clinical trials. *International Journal of Obesity*, *36*(12), 1485–1493.

Abete, I., Parra, D., & Martinez, J. A. (2009). Legume-, fish-, or high-protein-based hypocaloric diets: Effects on weight loss and mitochondrial oxidation in obese men. *Journal of Medicinal Food*, *12*, 100–108.

Abou Samra, R., Keersmaekers, L., Brienza, D., Mukherjee, R., & Mace, K. (2011). Effect of different protein sources on satiation and short-term satiety when consumed as a starter. *Nutrition Journal*, *10*, 139.

Acheson, K. J., Blondel-Lubrano, A., Oguey-Araymon, S., Beaumont, M., Emady-Azar, S., Ammon-Zufferey, C., et al. (2011). Protein choices targeting thermogenesis and metabolism. *The American Journal of Clinical Nutrition*, *93*, 525–534.

Acheson, K. J., Gremaud, G., Meirim, I., Montigon, F., Krebs, Y., Fay, L. B., et al. (2004). Metabolic effects of caffeine in humans: Lipid oxidation or futile cycling? *The American Journal of Clinical Nutrition*, *79*, 40–46.

Acheson, K. J., Zahorska-Markiewicz, B., Pittet, P., Anantharaman, K., & Jequier, E. (1980). Caffeine and coffee: Their influence on metabolic rate and substrate utilization in normal weight and obese individuals. *The American Journal of Clinical Nutrition*, *33*, 989–997.

Albert, C. M., Gaziano, J. M., Willett, W. C., & Manson, J. E. (2002). Nut consumption and decreased risk of sudden cardiac death in the Physicians' Health Study. *Archives of Internal Medicine*, *162*, 1382–1387.

Alinia, S., Hels, O., & Tetens, I. (2009). The potential association between fruit intake and body weight—A review. *Obesity Reviews*, *10*, 639–647.

Almiron-Roig, E., & Drewnowski, A. (2003). Hunger, thirst, and energy intakes following consumption of caloric beverages. *Physiology and Behavior*, *79*, 767–773.

Alper, C. M., & Mattes, R. D. (2002). Effects of chronic peanut consumption on energy balance and hedonics. *International Journal of Obesity and Related Metabolic Disorders*, *26*, 1129–1137.

Anderson, G. H., & Moore, S. E. (2004). Dietary proteins in the regulation of food intake and body weight in humans. *The Journal of Nutrition*, *134*, 974S–979S.

Andersson, R., Fransson, G., Tietjen, M., & Aman, P. (2009). Content and molecular-weight distribution of dietary fiber components in whole-grain rye flour and bread. *Journal of Agricultural and Food Chemistry, 57*, 2004–2008.

Appel, L. J., Moore, T. J., Obarzanek, E., Vollmer, W. M., Svetkey, L. P., Sacks, F. M., et al. (1997). A clinical trial of the effects of dietary patterns on blood pressure. DASH Collaborative Research Group. *The New England Journal of Medicine, 336*, 1117–1124.

Archer, B. J., Johnson, S. K., Devereux, H. M., & Baxter, A. L. (2004). Effect of fat replacement by inulin or lupin-kernel fibre on sausage patty acceptability, post-meal perceptions of satiety and food intake in men. *The British Journal of Nutrition, 91*, 591–599.

Arciero, P. J., Gardner, A. W., Calles-Escandon, J., Benowitz, N. L., & Poehlman, E. T. (1995). Effects of caffeine ingestion on NE kinetics, fat oxidation, and energy expenditure in younger and older men. *The American Journal of Physiology, 268*, E1192–E1198.

Aston, L. M., Stokes, C. S., & Jebb, S. A. (2008). No effect of a diet with a reduced glycaemic index on satiety, energy intake and body weight in overweight and obese women. *International Journal of Obesity, 32*, 160–165.

Astrup, A., Toubro, S., Cannon, S., Hein, P., Breum, L., & Madsen, J. (1990). Caffeine: A double-blind, placebo-controlled study of its thermogenic, metabolic, and cardiovascular effects in healthy volunteers. *The American Journal of Clinical Nutrition, 51*, 759–767.

Baer, D. J., Stote, K. S., Paul, D. R., Harris, G. K., Rumpler, W. V., & Clevidence, B. A. (2011). Whey protein but not soy protein supplementation alters body weight and composition in free-living overweight and obese adults. *The Journal of Nutrition, 141*, 1489–1494.

Batterham, R. L., Heffron, H., Kapoor, S., Chivers, J. E., Chandarana, K., Herzog, H., et al. (2006). Critical role for peptide YY in protein-mediated satiation and body-weight regulation. *Cell Metabolism, 4*, 223–233.

Bautista-Castano, I., & Serra-Majem, L. (2012). Relationship between bread consumption, body weight, and abdominal fat distribution: Evidence from epidemiological studies. *Nutrition Reviews, 70*, 218–233.

Bazzano, L. A., Song, Y., Bubes, V., Good, C. K., Manson, J. E., & Liu, S. (2005). Dietary intake of whole and refined grain breakfast cereals and weight gain in men. *Obesity Research, 13*, 1952–1960.

Beck, E. J., Tapsell, L. C., Batterham, M. J., Tosh, S. M., & Huang, X. F. (2009). Increases in peptide Y-Y levels following oat beta-glucan ingestion are dose-dependent in overweight adults. *Nutrition Research, 29*, 705–709.

Beck, E. J., Tosh, S. M., Batterham, M. J., Tapsell, L. C., & Huang, X. F. (2009). Oat beta-glucan increases postprandial cholecystokinin levels, decreases insulin response and extends subjective satiety in overweight subjects. *Molecular Nutrition and Food Research, 53*, 1343–1351.

Bell, E. A., Castellanos, V. H., Pelkman, C. L., Thorwart, M. L., & Rolls, B. J. (1998). Energy density of foods affects energy intake in normal-weight women. *The American Journal of Clinical Nutrition, 67*, 412–420.

Bell, E. A., & Rolls, B. J. (2001). Energy density of foods affects energy intake across multiple levels of fat content in lean and obese women. *The American Journal of Clinical Nutrition, 73*, 1010–1018.

Bes-Rastrollo, M., Sabate, J., Gomez-Gracia, E., Alonso, A., Martinez, J. A., & Martinez-Gonzalez, M. A. (2007). Nut consumption and weight gain in a Mediterranean cohort: The SUN study. *Obesity (Silver Spring), 15*, 107–116.

Blatt, A. D., Roe, L. S., & Rolls, B. J. (2011a). Hidden vegetables: An effective strategy to reduce energy intake and increase vegetable intake in adults. *The American Journal of Clinical Nutrition, 93*, 756–763.

Blatt, A. D., Roe, L. S., & Rolls, B. J. (2011b). Increasing the protein content of meals and its effect on daily energy intake. *Journal of the American Dietetic Association, 111*, 290–294.

Blundell, J. (2010). Making claims: Functional foods for managing appetite and weight. *Nature Reviews. Endocrinology, 6*, 53–56.

Blundell, J., de Graaf, C., Hulshof, T., Jebb, S., Livingstone, B., Lluch, A., et al. (2010). Appetite control: Methodological aspects of the evaluation of foods. *Obesity Reviews, 11*, 251–270.

Blundell, J. E. (1999). The control of appetite: Basic concepts and practical implications. *Schweizerische Medizinische Wochenschrift, 129*, 182–188.

Blundell, J. E., & Finlayson, G. (2004). Is susceptibility to weight gain characterized by homeostatic or hedonic risk factors for overconsumption? *Physiology and Behavior, 82*, 21–25.

Blundell, J. E., Levin, F., King, N. A., Barkeling, B., Gustafsson, T., Hellstrom, P. M., et al. (2008). Overconsumption and obesity: Peptides and susceptibility to weight gain. *Regulatory Peptides, 149*, 32–38.

Bodinham, C. L., Hitchen, K. L., Youngman, P. J., Frost, G. S., & Robertson, M. D. (2011). Short-term effects of whole-grain wheat on appetite and food intake in healthy adults: A pilot study. *The British Journal of Nutrition, 106*, 327–330.

Boirie, Y., Dangin, M., Gachon, P., Vasson, M. P., Maubois, J. L., & Beaufrere, B. (1997). Slow and fast dietary proteins differently modulate postprandial protein accretion. In: *Proceedings of the National Academy of Sciences of the United States of America, 94*, 14930–14935.

Bolton, R. P., Heaton, K. W., & Burroughs, L. F. (1981). The role of dietary fiber in satiety, glucose, and insulin: Studies with fruit and fruit juice. *The American Journal of Clinical Nutrition, 34*, 211–217.

Booth, D. A. (2008). Physiological regulation through learnt control of appetites by contingencies among signals from external and internal environments. *Appetite, 51*, 433–441.

Booth, D. A. (2009). Lines, dashed lines and "scale" ex-tricks. Objective measurements of appetite versus subjective tests of intake. *Appetite, 53*, 434–437.

Bornet, F. R., Jardy-Gennetier, A. E., Jacquet, N., & Stowell, J. (2007). Glycaemic response to foods: Impact on satiety and long-term weight regulation. *Appetite, 49*, 535–553.

Borzoei, S., Neovius, M., Barkeling, B., Teixeira-Pinto, A., & Rossner, S. (2006). A comparison of effects of fish and beef protein on satiety in normal weight men. *European Journal of Clinical Nutrition, 60*, 897–902.

Bowen, J., Noakes, M., Trenerry, C., & Clifton, P. M. (2006). Energy intake, ghrelin, and cholecystokinin after different carbohydrate and protein preloads in overweight men. *The Journal of Clinical Endocrinology and Metabolism, 91*, 1477–1483.

Bracco, D., Ferrarra, J. M., Arnaud, M. J., Jequier, E., & Schutz, Y. (1995). Effects of caffeine on energy metabolism, heart rate, and methylxanthine metabolism in lean and obese women. *The American Journal of Physiology, 269*, E671–E678.

Bray, G. A., Smith, S. R., de Jonge, L., Xie, H., Rood, J., Martin, C. K., et al. (2012). Effect of dietary protein content on weight gain, energy expenditure, and body composition during overeating: A randomized controlled trial. *Journal of the American Medical Association, 307*, 47–55.

Brennan, A. M., Sweeney, L. L., Liu, X., & Mantzoros, C. S. (2010). Walnut consumption increases satiation but has no effect on insulin resistance or the metabolic profile over a 4-day period. *Obesity (Silver Spring), 18*, 1176–1182.

Brown, A. (2008). *Understanding food, principles and preparation* (3rd ed.). Belmont, CA: Thomson Wadsworth, pp. 41–42.

Buijsse, B., Feskens, E. J., Schulze, M. B., Forouhi, N. G., Wareham, N. J., Sharp, S., et al. (2009). Fruit and vegetable intakes and subsequent changes in body weight in European populations: Results from the project on Diet, Obesity, and Genes (DiOGenes). *The American Journal of Clinical Nutrition, 90*, 202–209.

Bujnowski, D., Xun, P., Daviglus, M. L., Van Horn, L., He, K., & Stamler, J. (2011). Longitudinal association between animal and vegetable protein intake and obesity among men in the United States: The Chicago Western Electric Study. *Journal of the American Dietetic Association, 111*(1150–1155), e1151.

Burns, A. A., Livingstone, M. B., Welch, R. W., Dunne, A., Reid, C. A., & Rowland, I. R. (2001). The effects of yoghurt containing a novel fat emulsion on energy and macronutrient intakes in non-overweight, overweight and obese subjects. *International Journal of Obesity and Related Metabolic Disorders, 25*, 1487–1496.

Burns, A. A., Livingstone, M. B., Welch, R. W., Dunne, A., Robson, P. J., Lindmark, L., et al. (2000). Short-term effects of yoghurt containing a novel fat emulsion on energy and macronutrient intakes in non-obese subjects. *International Journal of Obesity and Related Metabolic Disorders, 24*, 1419–1425.

Burns, A. A., Livingstone, M. B., Welch, R. W., Dunne, A., & Rowland, I. R. (2002). Dose–response effects of a novel fat emulsion (Olibra) on energy and macronutrient intakes up to 36 h post-consumption. *European Journal of Clinical Nutrition, 56*, 368–377.

Canfi, A., Gepner, Y., Schwarzfuchs, D., Golan, R., Shahar, D. R., Fraser, D., et al. (2011). Effect of changes in the intake of weight of specific food groups on successful body weight loss during a multi-dietary strategy intervention trial. *Journal of the American College of Nutrition, 30*, 491–501.

Cardello, A. V., Schutz, H. G., Lesher, L. L., & Merrill, E. (2005). Development and testing of a labeled magnitude scale of perceived satiety. *Appetite, 44*, 1–13.

Casas-Agustench, P., Lopez-Uriarte, P., Bullo, M., Ros, E., Gomez-Flores, A., & Salas-Salvado, J. (2009). Acute effects of three high-fat meals with different fat saturations on energy expenditure, substrate oxidation and satiety. *Clinical Nutrition, 28*, 39–45.

Cassady, B. A., Hollis, J. H., Fulford, A. D., Considine, R. V., & Mattes, R. D. (2009). Mastication of almonds: Effects of lipid bioaccessibility, appetite, and hormone response. *The American Journal of Clinical Nutrition, 89*, 794–800.

Chan, Y. K., Strik, C. M., Budgett, S. C., McGill, A. T., Proctor, J., & Poppitt, S. D. (2012). The emulsified lipid Fabuless (Olibra) does not decrease food intake but suppresses appetite when consumed with yoghurt but not alone or with solid foods: A food effect study. *Physiology and Behavior, 105*, 742–748.

Chang, U. J., Hong, Y. H., Suh, H. J., & Jung, E. Y. (2010). Lowering the energy density of parboiled rice by adding water-rich vegetables can decrease total energy intake in a parboiled rice-based diet without reducing satiety on healthy women. *Appetite, 55*, 338–342.

Charlton, K. E., Tapsell, L. C., Batterham, M. J., Thorne, R., O'Shea, J., Zhang, Q., et al. (2011). Pork, beef and chicken have similar effects on acute satiety and hormonal markers of appetite. *Appetite, 56*, 1–8.

Chaudhri, O. B., Field, B. C., & Bloom, S. R. (2008). Gastrointestinal satiety signals. *International Journal of Obesity, 32*(Suppl. 7), S28–S31.

Cheskin, L. J., Davis, L. M., Lipsky, L. M., Mitola, A. H., Lycan, T., Mitchell, V., et al. (2008). Lack of energy compensation over 4 days when white button mushrooms are substituted for beef. *Appetite, 51*, 50–57.

Cho, S., Dietrich, M., Brown, C. J., Clark, C. A., & Block, G. (2003). The effect of breakfast type on total daily energy intake and body mass index: Results from the Third National Health and Nutrition Examination Survey (NHANES III). *Journal of the American College of Nutrition, 22*, 296–302.

Chu, B. S., Rich, G. T., Ridout, M. J., Faulks, R. M., Wickham, M. S., & Wilde, P. J. (2009). Modulating pancreatic lipase activity with galactolipids: Effects of emulsion interfacial composition. *Langmuir, 25*, 9352–9360.

Clegg, M., & Shafat, A. (2010). Energy and macronutrient composition of breakfast affect gastric emptying of lunch and subsequent food intake, satiety and satiation. *Appetite, 54*, 517–523.

Clegg, M. E., McKenna, P., McClean, C., Davison, G. W., Trinick, T., Duly, E., et al. (2011). Gastrointestinal transit, post-prandial lipaemia and satiety following 3 days high-fat diet in men. *European Journal of Clinical Nutrition, 65*, 240–246.

Cleveland, L. E., Moshfegh, A. J., Albertson, A. M., & Goldman, J. D. (2000). Dietary intake of whole grains. *Journal of the American College of Nutrition, 19*, 331S–338S.

Clifton, P. M., Keogh, J. B., & Noakes, M. (2008). Long-term effects of a high-protein weight-loss diet. *The American Journal of Clinical Nutrition, 87*, 23–29.

Cope, M. B., Erdman, J. W., Jr., & Allison, D. B. (2008). The potential role of soyfoods in weight and adiposity reduction: An evidence-based review. *Obesity Reviews, 9*, 219–235.

Crovetti, R., Porrini, M., Santangelo, A., & Testolin, G. (1998). The influence of thermic effect of food on satiety. *European Journal of Clinical Nutrition, 52*, 482–488.

Dangin, M., Boirie, Y., Guillet, C., & Beaufrere, B. (2002). Influence of the protein digestion rate on protein turnover in young and elderly subjects. *The Journal of Nutrition, 132*, 3228S–3233S.

Darzi, J., Frost, G. S., & Robertson, M. D. (2011). Do SCFA have a role in appetite regulation? *The Proceedings of the Nutrition Society, 70*, 119–128.

Das, S. K., Gilhooly, C. H., Golden, J. K., Pittas, A. G., Fuss, P. J., Cheatham, R. A., et al. (2007). Long-term effects of 2 energy-restricted diets differing in glycemic load on dietary adherence, body composition, and metabolism in CALERIE: A 1-y randomized controlled trial. *The American Journal of Clinical Nutrition, 85*, 1023–1030.

Dauthy, M. E., Food and Agriculture Organization of the United Nations (1995). Fruit and vegetable processing. FAO agricultural services bulletin no.119. Food and Agriculture Organization of the United Nations, Rome.

Davidi, A., Reynolds, J., Njike, V. Y., Ma, Y., Doughty, K., & Katz, D. L. (2011). The effect of the addition of daily fruit and nut bars to diet on weight, and cardiac risk profile, in overweight adults. *Journal of Human Nutrition and Dietetics, 24*, 543–551.

De Araujo, I. E., & Rolls, E. T. (2004). Representation in the human brain of food texture and oral fat. *The Journal of Neuroscience, 24*, 3086–3093.

de Graaf, C., Blom, W. A., Smeets, P. A., Stafleu, A., & Hendriks, H. F. (2004). Biomarkers of satiation and satiety. *The American Journal of Clinical Nutrition, 79*, 946–961.

de Oliveira, M. C., Sichieri, R., & Venturim Mozzer, R. (2008). A low-energy-dense diet adding fruit reduces weight and energy intake in women. *Appetite, 51*, 291–295.

DHHS, (2009). *FDA provides guidance on whole grains for manufacturers: US Food and Drug Administration, 2009.* http://www.fda.gov/NewsEvents/Newsroom/PressAnnouncements/2006/ucm108598.htm. Accessed on September 17, 2012.

Dhurandhar, N. V. (2012). When commonsense does not make sense. *International Journal of Obesity, 36*, 1332–1333.

Diepvens, K., Soenen, S., Steijns, J., Arnold, M., & Westerterp-Plantenga, M. (2007). Long-term effects of consumption of a novel fat emulsion in relation to body-weight management. *International Journal of Obesity, 31*, 942–949.

Diepvens, K., Steijns, J., Zuurendonk, P., & Westerterp-Plantenga, M. S. (2008). Short-term effects of a novel fat emulsion on appetite and food intake. *Physiology and Behavior, 95*, 114–117.

Dougkas, A., Minihane, A. M., Givens, D. I., Reynolds, C. K., & Yaqoob, P. (2012). Differential effects of dairy snacks on appetite, but not overall energy intake. *The British Journal of Nutrition, 108*, 2274–2285.

Dougkas, A., Reynolds, C. K., Givens, I. D., Elwood, P. C., & Minihane, A. M. (2011). Associations between dairy consumption and body weight: A review of the evidence and underlying mechanisms. *Nutrition Research Reviews, 24*, 72–95.

Dove, E. R., Hodgson, J. M., Puddey, I. B., Beilin, L. J., Lee, Y. P., & Mori, T. A. (2009). Skim milk compared with a fruit drink acutely reduces appetite and energy intake in overweight men and women. *The American Journal of Clinical Nutrition, 90*, 70–75.

Drapeau, V., Blundell, J., Therrien, F., Lawton, C., Richard, D., & Tremblay, A. (2005). Appetite sensations as a marker of overall intake. *The British Journal of Nutrition*, *93*, 273–280.

Drapeau, V., Despres, J. P., Bouchard, C., Allard, L., Fournier, G., Leblanc, C., et al. (2004). Modifications in food-group consumption are related to long-term body-weight changes. *The American Journal of Clinical Nutrition*, *80*, 29–37.

Drapeau, V., King, N., Hetherington, M., Doucet, E., Blundell, J., & Tremblay, A. (2007). Appetite sensations and satiety quotient: Predictors of energy intake and weight loss. *Appetite*, *48*, 159–166.

Drewnowski, A. (1997). Taste preferences and food intake. *Annual Review of Nutrition*, *17*, 237–253.

Drewnowski, A. (1998). Energy density, palatability, and satiety: Implications for weight control. *Nutrition Reviews*, *56*, 347–353.

Drewnowski, A. (2003). The role of energy density. *Lipids*, *38*, 109–115.

Drewnowski, A., & Almiron-Roig, E. (2010). Human perceptions and preferences for fat-rich foods. In J. P. Montmayeur & J. le Coutre (Eds.), *Fat detection: Taste, texture, and post ingestive effects*. Boca Raton, FL: CRC Press.

Du, H., van der, A. D., Boshuizen, H. C., Forouhi, N. G., Wareham, N. J., Halkjaer, J., et al. (2010). Dietary fiber and subsequent changes in body weight and waist circumference in European men and women. *The American Journal of Clinical Nutrition*, *91*, 329–336.

Dulloo, A. G., Geissler, C. A., Horton, T., Collins, A., & Miller, D. S. (1989). Normal caffeine consumption: Influence on thermogenesis and daily energy expenditure in lean and postobese human volunteers. *The American Journal of Clinical Nutrition*, *49*, 44–50.

Ebbeling, C. B., Leidig, M. M., Sinclair, K. B., Hangen, J. P., & Ludwig, D. S. (2003). A reduced-glycemic load diet in the treatment of adolescent obesity. *Archives of Pediatrics and Adolescent Medicine*, *157*, 773–779.

Ellis, P. R., Kendall, C. W., Ren, Y., Parker, C., Pacy, J. F., Waldron, K. W., et al. (2004). Role of cell walls in the bioaccessibility of lipids in almond seeds. *The American Journal of Clinical Nutrition*, *80*, 604–613.

Ello-Martin, J. A., Roe, L. S., Ledikwe, J. H., Beach, A. M., & Rolls, B. J. (2007). Dietary energy density in the treatment of obesity: A year-long trial comparing 2 weight-loss diets. *The American Journal of Clinical Nutrition*, *85*, 1465–1477.

Farajian, P., Katsagani, M., & Zampelas, A. (2010). Short-term effects of a snack including dried prunes on energy intake and satiety in normal-weight individuals. *Eating Behaviors*, *11*, 201–203.

Feltrin, K. L., Little, T. J., Meyer, J. H., Horowitz, M., Smout, A. J., Wishart, J., et al. (2004). Effects of intraduodenal fatty acids on appetite, antropyloroduodenal motility, and plasma CCK and GLP-1 in humans vary with their chain length. *American Journal of Physiology Regulatory, Integrative and Comparative Physiology*, *287*, R524–R533.

Field, B. C., Chaudhri, O. B., & Bloom, S. R. (2010). Bowels control brain: Gut hormones and obesity. *Nature Reviews Endocrinology*, *6*, 444–453.

Flegal, K. M., Carroll, M. D., Kit, B. K., & Ogden, C. L. (2012). Prevalence of obesity and trends in the distribution of body mass index among US adults, 1999–2010. *Journal of the American Medical Association*, *307*, 491–497.

Flint, A., Raben, A., Blundell, J. E., & Astrup, A. (2000). Reproducibility, power and validity of visual analogue scales in assessment of appetite sensations in single test meal studies. *International Journal of Obesity and Related Metabolic Disorders*, *24*, 38–48.

Flood-Obbagy, J. E., & Rolls, B. J. (2009). The effect of fruit in different forms on energy intake and satiety at a meal. *Appetite*, *52*, 416–422.

Flood, J. E., & Rolls, B. J. (2007). Soup preloads in a variety of forms reduce meal energy intake. *Appetite*, *49*, 626–634.

Fone, D. R., Horowitz, M., Read, N. W., Dent, J., & Maddox, A. (1990). The effect of terminal ileal triglyceride infusion on gastroduodenal motility and the intragastric distribution of a solid meal. *Gastroenterology, 98,* 568–575.

Foster, G. D., Shantz, K. L., Vander Veur, S. S., Oliver, T. L., Lent, M. R., Virus, A., et al. (2012). A randomized trial of the effects of an almond-enriched, hypocaloric diet in the treatment of obesity. *The American Journal of Clinical Nutrition, 96,* 249–254.

Fraser, G. E., Bennett, H. W., Jaceldo, K. B., & Sabate, J. (2002). Effect on body weight of a free 76 Kilojoule (320 calorie) daily supplement of almonds for six months. *Journal of the American College of Nutrition, 21,* 275–283.

Fraser, G. E., Sabate, J., Beeson, W. L., & Strahan, T. M. (1992). A possible protective effect of nut consumption on risk of coronary heart disease. The Adventist Health Study. *Archives of Internal Medicine, 152,* 1416–1424.

French, S. J., Conlon, C. A., Mutuma, S. T., Arnold, M., Read, N. W., Meijer, G., et al. (2000). The effects of intestinal infusion of long-chain fatty acids on food intake in humans. *Gastroenterology, 119,* 943–948.

French, S. J., Murray, B., Rumsey, R. D., Fadzlin, R., & Read, N. W. (1995). Adaptation to high-fat diets: Effects on eating behaviour and plasma cholecystokinin. *The British Journal of Nutrition, 73,* 179–189.

Froetschel, M. A. (1996). Bioactive peptides in digesta that regulate gastrointestinal function and intake. *Journal of Animal Science, 74,* 2500–2508.

Gaesser, G. A. (2007). Carbohydrate quantity and quality in relation to body mass index. *Journal of the American Dietetic Association, 107,* 1768–1780.

Gilbert, J. A., Joanisse, D. R., Chaput, J. P., Miegueu, P., Cianflone, K., Almeras, N., et al. (2011). Milk supplementation facilitates appetite control in obese women during weight loss: A randomised, single-blind, placebo-controlled trial. *The British Journal of Nutrition, 105,* 133–143.

Gilbertson, T. A., Fontenot, D. T., Liu, L., Zhang, H., & Monroe, W. T. (1997). Fatty acid modulation of K+ channels in taste receptor cells: Gustatory cues for dietary fat. *The American Journal of Physiology, 272,* C1203–C1210.

Gortmaker, S. L., Swinburn, B. A., Levy, D., Carter, R., Mabry, P. L., Finegood, D. T., et al. (2011). Changing the future of obesity: Science, policy, and action. *The Lancet, 378,* 838–847.

Gosnell, B. A., & Levine, A. S. (2009). Reward systems and food intake: Role of opioids. *International Journal of Obesity, 33*(Suppl. 2), S54–S58.

Greenway, F., O'Neil, C. E., Stewart, L., Rood, J., Keenan, M., & Martin, R. (2007). Fourteen weeks of treatment with Viscofiber increased fasting levels of glucagon-like peptide-1 and peptide-YY. *Journal of Medicinal Food, 10,* 720–724.

Greenwood, D. C., Cade, J. E., Draper, A., Barrett, J. H., Calvert, C., & Greenhalgh, A. (2000). Seven unique food consumption patterns identified among women in the UK Women's Cohort Study. *European Journal of Clinical Nutrition, 54,* 314–320.

Guh, D. P., Zhang, W., Bansback, N., Amarsi, Z., Birmingham, C. L., & Anis, A. H. (2009). The incidence of co-morbidities related to obesity and overweight: A systematic review and meta-analysis. *BMC Public Health, 9,* 88.

Gustafsson, K., Asp, N. G., Hagander, B., & Nyman, M. (1993). Effects of different vegetables in mixed meals on glucose homeostasis and satiety. *European Journal of Clinical Nutrition, 47,* 192–200.

Gustafsson, K., Asp, N. G., Hagander, B., & Nyman, M. (1994). Dose–response effects of boiled carrots and effects of carrots in lactic acid in mixed meals on glycaemic response and satiety. *European Journal of Clinical Nutrition, 48,* 386–396.

Gustafsson, K., Asp, N. G., Hagander, B., & Nyman, M. (1995). Satiety effects of spinach in mixed meals: Comparison with other vegetables. *International Journal of Food Sciences and Nutrition, 46,* 327–334.

Haber, G. B., Heaton, K. W., Murphy, D., & Burroughs, L. F. (1977). Depletion and disruption of dietary fibre. Effects on satiety, plasma-glucose, and serum-insulin. *The Lancet, 2*, 679–682.

Haenni, A., Sundberg, B., Yazdanpandah, N., Viberg, A., & Olsson, J. (2009). Effect of fat emulsion (Fabuless) on orocecal transit time in healthy men. *Scandinavian Journal of Gastroenterology, 44*, 1186–1190.

Halford, J. C., & Harrold, J. A. (2012). Satiety-enhancing products for appetite control: Science and regulation of functional foods for weight management. *The Proceedings of the Nutrition Society, 71*, 350–362.

Halkjaer, J., Olsen, A., Overvad, K., Jakobsen, M. U., Boeing, H., Buijsse, B., et al. (2011). Intake of total, animal and plant protein and subsequent changes in weight or waist circumference in European men and women: The Diogenes project. *International Journal of Obesity, 35*, 1104–1113.

Halkjaer, J., Tjonneland, A., Overvad, K., & Sorensen, T. I. (2009). Dietary predictors of 5-year changes in waist circumference. *Journal of the American Dietetic Association, 109*, 1356–1366.

Halkjaer, J., Tjonneland, A., Thomsen, B. L., Overvad, K., & Sorensen, T. I. (2006). Intake of macronutrients as predictors of 5-y changes in waist circumference. *The American Journal of Clinical Nutrition, 84*, 789–797.

Hall, W. L., Millward, D. J., Long, S. J., & Morgan, L. M. (2003). Casein and whey exert different effects on plasma amino acid profiles, gastrointestinal hormone secretion and appetite. *The British Journal of Nutrition, 89*, 239–248.

Hamedani, A., Akhavan, T., Abou Samra, R., & Anderson, G. H. (2009). Reduced energy intake at breakfast is not compensated for at lunch if a high-insoluble-fiber cereal replaces a low-fiber cereal. *American Journal of Clinical Nutrition, 89*, 1343–1349.

Haque, E., Chand, R., & Kapila, S. (2009). Biofunctional properties of bioactive peptides of milk origin. *Food Reviews International, 25*, 28–43.

Harper, A., James, A., Flint, A., & Astrup, A. (2007). Increased satiety after intake of a chocolate milk drink compared with a carbonated beverage, but no difference in subsequent ad libitum lunch intake. *The British Journal of Nutrition, 97*, 579–583.

Harris, K. A., & Kris-Etherton, P. M. (2010). Effects of whole grains on coronary heart disease risk. *Current Atherosclerosis Reports, 12*, 368–376.

He, K., Hu, F. B., Colditz, G. A., Manson, J. E., Willett, W. C., & Liu, S. (2004). Changes in intake of fruits and vegetables in relation to risk of obesity and weight gain among middle-aged women. *International Journal of Obesity and Related Metabolic Disorders, 28*, 1569–1574.

Hermsdorff, H. H., Zulet, M. A., Abete, I., & Martinez, J. A. (2011). A legume-based hypocaloric diet reduces proinflammatory status and improves metabolic features in overweight/obese subjects. *European Journal of Nutrition, 50*, 61–69.

Hess, J. R., Birkett, A. M., Thomas, W., & Slavin, J. L. (2011). Effects of short-chain fructooligosaccharides on satiety responses in healthy men and women. *Appetite, 56*, 128–134.

Hession, M., Rolland, C., Kulkarni, U., Wise, A., & Broom, J. (2009). Systematic review of randomized controlled trials of low-carbohydrate vs. low-fat/low-calorie diets in the management of obesity and its comorbidities. *Obesity Reviews, 10*, 36–50.

Hlebowicz, J., Darwiche, G., Bjorgell, O., & Almer, L. O. (2008). Effect of muesli with 4 g oat beta-glucan on postprandial blood glucose, gastric emptying and satiety in healthy subjects: A randomized crossover trial. *Journal of the American College of Nutrition, 27*, 470–475.

Hlebowicz, J., Wickenberg, J., Fahlstrom, R., Bjorgell, O., Almer, L. O., & Darwiche, G. (2007). Effect of commercial breakfast fibre cereals compared with corn flakes on postprandial blood glucose, gastric emptying and satiety in healthy subjects: A randomized blinded crossover trial. *Nutrition Journal, 6*, 22.

Hodgson, J. M., Lee, Y. P., Puddey, I. B., Sipsas, S., Ackland, T. R., Beilin, L. J., et al. (2010). Effects of increasing dietary protein and fibre intake with lupin on body weight and composition and blood lipids in overweight men and women. *International Journal of Obesity, 34,* 1086–1094.

Hollands, M. A., Arch, J. R., & Cawthorne, M. A. (1981). A simple apparatus for comparative measurements of energy expenditure in human subjects: The thermic effect of caffeine. *The American Journal of Clinical Nutrition, 34,* 2291–2294.

Hollis, J., & Mattes, R. (2007a). Effect of chronic consumption of almonds on body weight in healthy humans. *The British Journal of Nutrition, 98,* 651–656.

Hollis, J. H., & Mattes, R. D. (2007b). Effect of increased dairy consumption on appetitive ratings and food intake. *Obesity (Silver Spring), 15,* 1520–1526.

Holt, S. H., Delargy, H. J., Lawton, C. L., & Blundell, J. E. (1999). The effects of high-carbohydrate vs high-fat breakfasts on feelings of fullness and alertness, and subsequent food intake. *International Journal of Food Sciences and Nutrition, 50,* 13–28.

Holt, S. H., Miller, J. C., Petocz, P., & Farmakalidis, E. (1995). A satiety index of common foods. *European Journal of Clinical Nutrition, 49,* 675–690.

Honselman, C. S., Painter, J. E., Kennedy-Hagan, K. J., Halvorson, A., Rhodes, K., Brooks, T. L., et al. (2011). In-shell pistachio nuts reduce caloric intake compared to shelled nuts. *Appetite, 57,* 414–417.

Howarth, N. C., Saltzman, E., & Roberts, S. B. (2001). Dietary fiber and weight regulation. *Nutrition Reviews, 59,* 129–139.

Hu, F. B., Stampfer, M. J., Manson, J. E., Rimm, E. B., Colditz, G. A., Rosner, B. A., et al. (1998). Frequent nut consumption and risk of coronary heart disease in women: Prospective cohort study. *BMJ, 317,* 1341–1345.

Hughes, G. M., Boyland, E. J., Williams, N. J., Mennen, L., Scott, C., Kirkham, T. C., et al. (2008). The effect of Korean pine nut oil (PinnoThin) on food intake, feeding behaviour and appetite: A double-blind placebo-controlled trial. *Lipids in Health and Disease, 7,* 6.

Hursel, R., Viechtbauer, W., & Westerterp-Plantenga, M. S. (2009). The effects of green tea on weight loss and weight maintenance: A meta-analysis. *International Journal of Obesity, 33,* 956–961.

Hursel, R., & Westerterp-Plantenga, M. S. (2010). Thermogenic ingredients and body weight regulation. *International Journal of Obesity, 34,* 659–669.

Isaksson, H., Fredriksson, H., Andersson, R., Olsson, J., & Aman, P. (2009). Effect of rye bread breakfasts on subjective hunger and satiety: A randomized controlled trial. *Nutrition Journal, 8,* 39.

Isaksson, H., Rakha, A., Andersson, R., Fredriksson, H., Olsson, J., & Aman, P. (2011). Rye kernel breakfast increases satiety in the afternoon—An effect of food structure. *Nutrition Journal, 10,* 31.

Isaksson, H., Sundberg, B., Aman, P., Fredriksson, H., & Olsson, J. (2008). Whole grain rye porridge breakfast improves satiety compared to refined wheat bread breakfast. *Food Nutrition Research, 52,* 1–7.

Isaksson, H., Tillander, I., Andersson, R., Olsson, J., Fredriksson, H., Webb, D. L., et al. (2012). Whole grain rye breakfast—Sustained satiety during three weeks of regular consumption. *Physiology and Behavior, 105,* 877–884.

Jacobs, D. R., Jr., Meyer, K. A., Kushi, L. H., & Folsom, A. R. (1998). Whole-grain intake may reduce the risk of ischemic heart disease death in postmenopausal women: The Iowa Women's Health Study. *The American Journal of Clinical Nutrition, 68,* 248–257.

Jakobsen, M. U., Due, K. M., Dethlefsen, C., Halkjaer, J., Holst, C., Forouhi, N. G., et al. (2011). Fish consumption does not prevent increase in waist circumference in European women and men. *The British Journal of Nutrition, 108,* 924–931.

Jenkins, D. J., Kendall, C. W., Augustin, L. S., Franceschi, S., Hamidi, M., Marchie, A., et al. (2002). Glycemic index: Overview of implications in health and disease. *The American Journal of Clinical Nutrition, 76*, 266S–273S.

Jenkins, D. J., Wolever, T. M., Taylor, R. H., Barker, H., Fielden, H., Baldwin, J. M., et al. (1981). Glycemic index of foods: A physiological basis for carbohydrate exchange. *The American Journal of Clinical Nutrition, 34*, 362–366.

Johansson, L., Tuomainen, P., Anttila, H., Rita, H., & Virkki, L. (2006). Effect of processing on the extractability of oat beta-glucan. *Food Chemistry, 105*, 1439–1445.

Johnson, I. T., & Gee, J. M. (1981). Effect of gel-forming gums on the intestinal unstirred layer and sugar transport in vitro. *Gut, 22*, 398–403.

Johnson, S. K., Thomas, S. J., & Hall, R. S. (2005). Palatability and glucose, insulin and satiety responses of chickpea flour and extruded chickpea flour bread eaten as part of a breakfast. *European Journal of Clinical Nutrition, 59*, 169–176.

Johnson, S. L., McPhee, L., & Birch, L. L. (1991). Conditioned preferences: Young children prefer flavors associated with high dietary fat. *Physiology and Behavior, 50*, 1245–1251.

Karlstrom, B., Vessby, B., Asp, N. G., Boberg, M., Lithell, H., & Berne, C. (1987). Effects of leguminous seeds in a mixed diet in non-insulin-dependent diabetic patients. *Diabetes Research, 5*, 199–205.

Karra, E., Chandarana, K., & Batterham, R. L. (2009). The role of peptide YY in appetite regulation and obesity. *The Journal of Physiology, 587*, 19–25.

Keast, D. R., O'Neil, C. E., & Jones, J. M. (2011). Dried fruit consumption is associated with improved diet quality and reduced obesity in US adults: National Health and Nutrition Examination Survey, 1999–2004. *Nutrition Research, 31*, 460–467.

Kennedy-Hagan, K., Painter, J. E., Honselman, C., Halvorson, A., Rhodes, K., & Skwir, K. (2011). The effect of pistachio shells as a visual cue in reducing caloric consumption. *Appetite, 57*, 418–420.

Keogh, J. B., Lau, C. W., Noakes, M., Bowen, J., & Clifton, P. M. (2007). Effects of meals with high soluble fibre, high amylose barley variant on glucose, insulin, satiety and thermic effect of food in healthy lean women. *European Journal of Clinical Nutrition, 61*, 597–604.

Kim, H., Behall, K. M., Vinyard, B., & Conway, J. M. (2006). Short-term satiety and glycemic response after consumption of whole grains with various amounts of beta-glucan. *Cereal Food World, 51*, 29–33.

Knight, E. L., Stampfer, M. J., Hankinson, S. E., Spiegelman, D., & Curhan, G. C. (2003). The impact of protein intake on renal function decline in women with normal function or mild renal insufficiency. *Annals of Internal Medicine, 138*, 460–467.

Knutson, L., Koenders, D. J., Fridblom, H., Viberg, A., Sein, A., & Lennernas, H. (2010). Gastrointestinal metabolism of a vegetable-oil emulsion in healthy subjects. *The American Journal of Clinical Nutrition, 92*, 515–524.

Koh-Banerjee, P., Franz, M., Sampson, L., Liu, S., Jacobs, D. R., Jr., Spiegelman, D., et al. (2004). Changes in whole-grain, bran, and cereal fiber consumption in relation to 8-y weight gain among men. *The American Journal of Clinical Nutrition, 80*, 1237–1245.

Kovacs, E. M., & Mela, D. J. (2006). Metabolically active functional food ingredients for weight control. *Obesity Reviews, 7*, 59–78.

Krieger, J. W., Sitren, H. S., Daniels, M. J., & Langkamp-Henken, B. (2006). Effects of variation in protein and carbohydrate intake on body mass and composition during energy restriction: A meta-regression 1. *The American Journal of Clinical Nutrition, 83*, 260–274.

Kristensen, M., & Jensen, M. G. (2011). Dietary fibres in the regulation of appetite and food intake. Importance of viscosity. *Appetite, 56*, 65–70.

Krog-Mikkelsen, I., Sloth, B., Dimitrov, D., Tetens, I., Bjorck, I., Flint, A., et al. (2011). A low glycemic index diet does not affect postprandial energy metabolism but decreases postprandial insulinemia and increases fullness ratings in healthy women. *The Journal of Nutrition, 141*, 1679–1684.

Lanou, A. J., & Barnard, N. D. (2008). Dairy and weight loss hypothesis: An evaluation of the clinical trials. *Nutrition Reviews, 66*, 272–279.

Latner, J. D., Rosewall, J. K., & Chisholm, A. M. (2008). Energy density effects on food intake, appetite ratings, and loss of control in women with binge eating disorder and weight-matched controls. *Eating Behaviors, 9*, 257–266.

Lattimer, J. M., & Haub, M. D. (2010). Effects of dietary fiber and its components on metabolic health. *Nutrients, 2*, 1266–1289.

Leathwood, P., & Pollet, P. (1988). Effects of slow release carbohydrates in the form of bean flakes on the evolution of hunger and satiety in man. *Appetite, 10*, 1–11.

Ledoux, T. A., Hingle, M. D., & Baranowski, T. (2011). Relationship of fruit and vegetable intake with adiposity: A systematic review. *Obesity Reviews, 12*, e143–e150.

Lee, Y. P., Mori, T. A., Sipsas, S., Barden, A., Puddey, I. B., Burke, V., et al. (2006). Lupin-enriched bread increases satiety and reduces energy intake acutely. *The American Journal of Clinical Nutrition, 84*, 975–980.

Lejeune, M. P., Kovacs, E. M., & Westerterp-Plantenga, M. S. (2003). Effect of capsaicin on substrate oxidation and weight maintenance after modest body-weight loss in human subjects. *The British Journal of Nutrition, 90*, 651–659.

Lejeune, M. P., Westerterp, K. R., Adam, T. C., Luscombe-Marsh, N. D., & Westerterp-Plantenga, M. S. (2006). Ghrelin and glucagon-like peptide 1 concentrations, 24-h satiety, and energy and substrate metabolism during a high-protein diet and measured in a respiration chamber. *The American Journal of Clinical Nutrition, 83*, 89–94.

Li, Z., Song, R., Nguyen, C., Zerlin, A., Karp, H., Naowamondhol, K., et al. (2010). Pistachio nuts reduce triglycerides and body weight by comparison to refined carbohydrate snack in obese subjects on a 12-week weight loss program. *Journal of the American College of Nutrition, 29*, 198–203.

Little, T. J., & Feinle-Bisset, C. (2011). Effects of dietary fat on appetite and energy intake in health and obesity—Oral and gastrointestinal sensory contributions. *Physiology and Behavior, 104*, 613–620.

Little, T. J., Horowitz, M., & Feinle-Bisset, C. (2007). Modulation by high-fat diets of gastrointestinal function and hormones associated with the regulation of energy intake: Implications for the pathophysiology of obesity. *The American Journal of Clinical Nutrition, 86*, 531–541.

Liu, S., Willett, W. C., Manson, J. E., Hu, F. B., Rosner, B., & Colditz, G. (2003). Relation between changes in intakes of dietary fiber and grain products and changes in weight and development of obesity among middle-aged women. *The American Journal of Clinical Nutrition, 78*, 920–927.

Livesey, G. (2005). Low-glycaemic diets and health: Implications for obesity. *The Proceedings of the Nutrition Society, 64*, 105–113.

Livesey, G., Taylor, R., Hulshof, T., & Howlett, J. (2008). Glycemic response and health–a systematic review and meta-analysis: Relations between dietary glycemic properties and health outcomes. *The American Journal of Clinical Nutrition, 87*, 258S–268S.

Logan, C. M., McCaffrey, T. A., Wallace, J. M., Robson, P. J., Welch, R. W., Dunne, A., et al. (2006). Investigation of the medium-term effects of Olibra trade mark fat emulsion on food intake in non-obese subjects. *European Journal of Clinical Nutrition, 60*, 1081–1091.

Lopez-Garcia, E., van Dam, R. M., Rajpathak, S., Willett, W. C., Manson, J. E., & Hu, F. B. (2006). Changes in caffeine intake and long-term weight change in men and women. *The American Journal of Clinical Nutrition, 83*, 674–680.

Lorenzen, J., Frederiksen, R., Hoppe, C., Hvid, R., & Astrup, A. (2012). The effect of milk proteins on appetite regulation and diet-induced thermogenesis. *European Journal of Clinical Nutrition, 66*, 622–627.

Lorenzen, J. K., Nielsen, S., Holst, J. J., Tetens, I., Rehfeld, J. F., & Astrup, A. (2007). Effect of dairy calcium or supplementary calcium intake on postprandial fat metabolism, appetite, and subsequent energy intake. *The American Journal of Clinical Nutrition, 85*, 678–687.

Louie, J. C., Flood, V. M., Hector, D. J., Rangan, A. M., & Gill, T. P. (2011). Dairy consumption and overweight and obesity: A systematic review of prospective cohort studies. *Obesity Reviews, 12*, e582–e592.

Ludwig, D. S. (2002). The glycemic index: Physiological mechanisms relating to obesity, diabetes, and cardiovascular disease. *Journal of the American Medical Association, 287*, 2414–2423.

Ludwig, D. S. (2003). Dietary glycemic index and the regulation of body weight. *Lipids, 38*, 117–121.

Ludwig, D. S., Pereira, M. A., Kroenke, C. H., Hilner, J. E., Van Horn, L., Slattery, M. L., et al. (1999). Dietary fiber, weight gain, and cardiovascular disease risk factors in young adults. *Journal of the American Medical Association, 282*, 1539–1546.

Luhovyy, B. L., Akhavan, T., & Anderson, G. H. (2007). Whey proteins in the regulation of food intake and satiety. *Journal of the American College of Nutrition, 26*, 704S–712S.

Lyly, M., Liukkonen, K. H., Salmenkallio-Marttila, M., Karhunen, L., Poutanen, K., & Lahteenmaki, L. (2009). Fibre in beverages can enhance perceived satiety. *European Journal of Nutrition, 48*, 251–258.

Major, G. C., Alarie, F. P., Dore, J., & Tremblay, A. (2009). Calcium plus vitamin D supplementation and fat mass loss in female very low-calcium consumers: Potential link with a calcium-specific appetite control. *The British Journal of Nutrition, 101*, 659–663.

Major, G. C., Chaput, J. P., Ledoux, M., St-Pierre, S., Anderson, G. H., Zemel, M. B., et al. (2008). Recent developments in calcium-related obesity research. *Obesity Reviews, 9*, 428–445.

Maki, K. C., Rains, T. M., Kaden, V. N., Raneri, K. R., & Davidson, M. H. (2007). Effects of a reduced-glycemic-load diet on body weight, body composition, and cardiovascular disease risk markers in overweight and obese adults. *The American Journal of Clinical Nutrition, 85*, 724–734.

Malik, V. S., Popkin, B. M., Bray, G. A., Despres, J. P., & Hu, F. B. (2010). Sugar-sweetened beverages, obesity, type 2 diabetes mellitus, and cardiovascular disease risk. *Circulation, 121*, 1356–1364.

Maljaars, J., Peters, H. P., & Masclee, A. M. (2007). Review article: The gastrointestinal tract: Neuroendocrine regulation of satiety and food intake. *Alimentary Pharmacology and Therapeutics, 26*(Suppl. 2), 241–250.

Maljaars, J., Romeyn, E. A., Haddeman, E., Peters, H. P., & Masclee, A. A. (2009). Effect of fat saturation on satiety, hormone release, and food intake. *The American Journal of Clinical Nutrition, 89*, 1019–1024.

Maljaars, P. W., Peters, H. P., Kodde, A., Geraedts, M., Troost, F. J., Haddeman, E., et al. (2011). Length and site of the small intestine exposed to fat influences hunger and food intake. *The British Journal of Nutrition, 106*, 1609–1615.

Maljaars, P. W., Peters, H. P., Mela, D. J., & Masclee, A. A. (2008). Ileal brake: A sensible food target for appetite control. A review. *Physiology and Behavior, 95*, 271–281.

Malkki, Y., & Virtanen, E. (2001). Gastrointestinal effects of oat bran and oat gum: A review. *Lebensmittel Wissenschaft und Technologie, 34*, 337–347.

Mancino, L., Kuchler, F., & Leibtag, E. (2008). Getting consumers to eat more whole-grains: The role of policy, information, and food manufacturers. *Food Policy, 33*, 489–496.

Marciani, L., Gowland, P. A., Spiller, R. C., Manoj, P., Moore, R. J., Young, P., et al. (2001). Effect of meal viscosity and nutrients on satiety, intragastric dilution, and emptying assessed by MRI. *American Journal of Physiology Gastrointestinal and Liver Physiology, 280*, G1227–G1233.

Martinez-Gonzalez, M. A., & Bes-Rastrollo, M. (2011). Nut consumption, weight gain and obesity: Epidemiological evidence. *Nutrition, Metabolism, and Cardiovascular Diseases, 21*(Suppl. 1), S40–S45.

Mattes, R. D. (2002). Ready-to-eat cereal used as a meal replacement promotes weight loss in humans. *Journal of the American College of Nutrition, 21*, 570–577.

Mattes, R. D. (2006). Fluid energy—Where's the problem? *Journal of the American Dietetic Association, 106*, 1956–1961.

Mattes, R. D. (2008). The energetics of nut consumption. *Asia Pacific Journal of Clinical Nutrition, 17*(Suppl. 1), 337–339.

Mattes, R. D. (2009). Oral detection of short-, medium-, and long-chain free fatty acids in humans. *Chemical Senses, 34*, 145–150.

Mattes, R. D., & Campbell, W. W. (2009). Effects of food form and timing of ingestion on appetite and energy intake in lean young adults and in young adults with obesity. *Journal of the American Dietetic Association, 109*, 430–437.

Mattes, R. D., Kris-Etherton, P. M., & Foster, G. D. (2008). Impact of peanuts and tree nuts on body weight and healthy weight loss in adults. *The Journal of Nutrition, 138*, 1741S–1745S.

Matthews, A., Hull, S., Angus, F., & Johnston, K. L. (2012). The effect of ready-to-eat cereal consumption on energy intake, body weight and anthropometric measurements: Results from a randomized, controlled intervention trial. *International Journal of Food Sciences and Nutrition, 63*, 107–113.

McAllister, E. J., Dhurandhar, N. V., Keith, S. W., Aronne, L. J., Barger, J., Baskin, M., et al. (2009). Ten putative contributors to the obesity epidemic. *Critical Reviews in Food Science and Nutrition, 49*, 868–913.

McCrory, M. A., Hamaker, B. R., Lovejoy, J. C., & Eichelsdoerfer, P. E. (2010). Pulse consumption, satiety, and weight management. *Advances in Nutrition, 1*, 17–30.

McCrory, M. A., Lovejoy, J. C., Palmer, M. A., Eichelsdoerfer, P. E., Gehrke, M. M., Kavanaugh, I. T., et al. (2008). Effectiveness of legume consumption for facilitating weight loss: A randomized trial (Abstract). *The FASEB Journal, 22*, 1084–1089.

Mela, D. J. (2006). Eating for pleasure or just wanting to eat? Reconsidering sensory hedonic responses as a driver of obesity. *Appetite, 47*, 10–17.

Melanson, K., Gootman, J., Myrdal, A., Kline, G., & Rippe, J. M. (2003). Weight loss and total lipid profile changes in overweight women consuming beef or chicken as the primary protein source. *Nutrition, 19*, 409–414.

Millward, D. J., Layman, D. K., Tome, D., & Schaafsma, G. (2008). Protein quality assessment: Impact of expanding understanding of protein and amino acid needs for optimal health. *The American Journal of Clinical Nutrition, 87*, 1576S–1581S.

Mollard, R. C., Wong, C. L., Luhovyy, B. L., & Anderson, G. H. (2011). First and second meal effects of pulses on blood glucose, appetite, and food intake at a later meal. *Applied Physiology, Nutrition and Metabolism, 36*, 634–642.

Mollard, R. C., Zykus, A., Luhovyy, B. L., Nunez, M. F., Wong, C. L., & Anderson, G. H. (2011). The acute effects of a pulse-containing meal on glycaemic responses and measures of satiety and satiation within and at a later meal. *The British Journal of Nutrition, 108*, 509–517.

Mori, A. M., Considine, R. V., & Mattes, R. D. (2011). Acute and second-meal effects of almond form in impaired glucose tolerant adults: A randomized crossover trial. *Nutrition and Metabolism, 8*, 6.

Morrison, C. D., Xi, X., White, C. L., Ye, J., & Martin, R. J. (2007). Amino acids inhibit Agrp gene expression via an mTOR-dependent mechanism. *American Journal of Physiology, Endocrinology and Metabolism, 293*, E165–E171.

Mozaffarian, D., Hao, T., Rimm, E. B., Willett, W. C., & Hu, F. B. (2011). Changes in diet and lifestyle and long-term weight gain in women and men. *The New England Journal of Medicine, 364,* 2392–2404.

Murty, C. M., Pittaway, J. K., & Ball, M. J. (2010). Chickpea supplementation in an Australian diet affects food choice, satiety and bowel health. *Appetite, 54,* 282–288.

NAP, (2005). *Dietary reference intakes for energy, carbohydrate, fiber, fat, fatty acids, cholesterol, protein, and amino acids (macronutrients). Food and Nutrition Board.* Washington, DC: The National Academies Press. http://www.nap.edu/openbook.php?isbn=0309085373. Accessed on September 17, 2012.

Nathan, P. J., & Bullmore, E. T. (2009). From taste hedonics to motivational drive: Central mu-opioid receptors and binge-eating behaviour. *The International Journal of Neuropsychopharmacology, 12,* 995–1008.

Natoli, S., & McCoy, P. (2007). A review of the evidence: Nuts and body weight. *Asia Pacific Journal of Clinical Nutrition, 16,* 588–597.

Nilsson, M., Stenberg, M., Frid, A. H., Holst, J. J., & Bjorck, I. M. (2004). Glycemia and insulinemia in healthy subjects after lactose-equivalent meals of milk and other food proteins: The role of plasma amino acids and incretins. *The American Journal of Clinical Nutrition, 80,* 1246–1253.

O'Neil, C. E., Zanovec, M., Cho, S. S., & Nicklas, T. A. (2010). Whole grain and fiber consumption are associated with lower body weight measures in US adults: National Health and Nutrition Examination Survey 1999–2004. *Nutrition Research, 30,* 815–822.

Ogden, C. L., Carroll, M. D., Kit, B. K., & Flegal, K. M. (2012). Prevalence of obesity and trends in body mass index among US children and adolescents, 1999–2010. *Journal of the American Medical Association, 307,* 483–490.

Olsson, J., Sundberg, B., Viberg, A., & Haenni, A. (2011). Effect of a vegetable-oil emulsion on body composition; a 12-week study in overweight women on a meal replacement therapy after an initial weight loss: A randomized controlled trial. *European Journal of Nutrition, 50,* 235–242.

Pal, S., & Ellis, V. (2010). The acute effects of four protein meals on insulin, glucose, appetite and energy intake in lean men. *The British Journal of Nutrition, 104,* 1241–1248.

Papanikolaou, Y., & Fulgoni, V. L., 3rd. (2008). Bean consumption is associated with greater nutrient intake, reduced systolic blood pressure, lower body weight, and a smaller waist circumference in adults: Results from the National Health and Nutrition Examination Survey 1999–2002. *Journal of the American College of Nutrition, 27,* 569–576.

Papathanasopoulos, A., & Camilleri, M. (2010). Dietary fiber supplements: Effects in obesity and metabolic syndrome and relationship to gastrointestinal functions. *Gastroenterology, 138*(65–72), e61–e62.

Pasman, W. J., Heimerikx, J., Rubingh, C. M., van den Berg, R., O'Shea, M., Gambelli, L., et al. (2008). The effect of Korean pine nut oil on in vitro CCK release, on appetite sensations and on gut hormones in post-menopausal overweight women. *Lipids in Health and Disease, 7,* 10.

Pasman, W. J., Westerterp-Plantenga, M. S., & Saris, W. H. (1997). The effectiveness of long-term supplementation of carbohydrate, chromium, fibre and caffeine on weight maintenance. *International Journal of Obesity and Related Metabolic Disorders, 21,* 1143–1151.

Perez-Escamilla, R., Obbagy, J. E., Altman, J. M., Essery, E. V., McGrane, M. M., Wong, Y. P., et al. (2012). Dietary energy density and body weight in adults and children: A systematic review. *Journal of the Academy of Nutrition and Dietetics, 112,* 671–684.

Peters, H., Beglinger, C., Mela, D., & Schuring, E. (2010). Comment and reply on: Effect of fat emulsion (Fabuless) on orocecal transit time in healthy men. *Scandinavian Journal of Gastroenterology, 45,* 637–638 author reply 638–639.

Peters, H. P., Boers, H. M., Haddeman, E., Melnikov, S. M., & Qvyjt, F. (2009). No effect of added beta-glucan or of fructooligosaccharide on appetite or energy intake. *The American Journal of Clinical Nutrition, 89*, 58–63.

Phung, O. J., Baker, W. L., Matthews, L. J., Lanosa, M., Thorne, A., & Coleman, C. I. (2010). Effect of green tea catechins with or without caffeine on anthropometric measures: A systematic review and meta-analysis. *The American Journal of Clinical Nutrition, 91*, 73–81.

Pironi, L., Stanghellini, V., Miglioli, M., Corinaldesi, R., De Giorgio, R., Ruggeri, E., et al. (1993). Fat-induced ileal brake in humans: A dose-dependent phenomenon correlated to the plasma levels of peptide YY. *Gastroenterology, 105*, 733–739.

Pombo-Rodrigues, S., Calame, W., & Re, R. (2011). The effects of consuming eggs for lunch on satiety and subsequent food intake. *International Journal of Food Sciences and Nutrition, 62*, 593–599.

Poppitt, S. D., Strik, C. M., MacGibbon, A. K., McArdle, B. H., Budgett, S. C., & McGill, A. T. (2010). Fatty acid chain length, postprandial satiety and food intake in lean men. *Physiology and Behavior, 101*, 161–167.

Potier, M., Fromentin, G., Calvez, J., Benamouzig, R., Martin-Rouas, C., Pichon, L., et al. (2009). A high-protein, moderate-energy, regular cheesy snack is energetically compensated in human subjects. *The British Journal of Nutrition, 102*, 625–631.

Powley, T. L., & Phillips, R. J. (2004). Gastric satiation is volumetric, intestinal satiation is nutritive. *Physiology and Behavior, 82*, 69–74.

Preston, S. H., & Stokes, A. (2011). Contribution of obesity to international differences in life expectancy. *American Journal of Public Health, 101*, 2137–2143.

Ragaee, S. M., Campbell, G. L., Scoles, G. J., McLeod, J. G., & Tyler, R. T. (2001). Studies on rye (Secale cereale L.) lines exhibiting a range of extract viscosities. 1. Composition, molecular weight distribution of water extracts, and biochemical characteristics of purified water-extractable arabinoxylan. *Journal of Agricultural and Food Chemistry, 49*, 2437–2445.

Rajaram, S., & Sabate, J. (2006). Nuts, body weight and insulin resistance. *The British Journal of Nutrition, 96*(Suppl. 2), S79–S86.

Ratliff, J., Leite, J. O., de Ogburn, R., Puglisi, M. J., VanHeest, J., & Fernandez, M. L. (2010). Consuming eggs for breakfast influences plasma glucose and ghrelin, while reducing energy intake during the next 24 hours in adult men. *Nutrition Research, 30*, 96–103.

Raynor, H. A., Looney, S. M., Steeves, E. A., Spence, M., & Gorin, A. A. (2012). The effects of an energy density prescription on diet quality and weight loss: A pilot randomized controlled trial. *Journal of the Academy of Nutrition and Dietetics, 112*, 1397–1402.

Read, N. W., McFarlane, A., Kinsman, R. I., Bates, T. E., Blackhall, N. W., Farrar, G. B., et al. (1984). Effect of infusion of nutrient solutions into the ileum on gastrointestinal transit and plasma levels of neurotensin and enteroglucagon. *Gastroenterology, 86*, 274–280.

Rebello, C. J., Martin, C. K., Johnson, W. D., O'Neil, C. E., & Greenway, F. L. (2012). Efficacy of Olibra: A 12-week randomized controlled trial and a review of earlier studies. *Journal of Diabetes Science and Technology, 6*, 695–708.

Rodriguez-Rodriguez, E., Lopez-Sobaler, A. M., Navarro, A. R., Bermejo, L. M., Ortega, R. M., & Andres, P. (2008). Vitamin B6 status improves in overweight/obese women following a hypocaloric diet rich in breakfast cereals, and may help in maintaining fat-free mass. *International Journal of Obesity, 32*, 1552–1558.

Roe, L. S., Meengs, J. S., & Rolls, B. J. (2012). Salad and satiety. The effect of timing of salad consumption on meal energy intake. *Appetite, 58*, 242–248.

Rokholm, B., Baker, J. L., & Sorensen, T. I. (2010). The levelling off of the obesity epidemic since the year 1999—A review of evidence and perspectives. *Obesity Reviews, 11*, 835–846.

Rolls, B. J., Bell, E. A., Castellanos, V. H., Chow, M., Pelkman, C. L., & Thorwart, M. L. (1999). Energy density but not fat content of foods affected energy intake in lean and obese women. *The American Journal of Clinical Nutrition, 69*, 863–871.

Rolls, B. J., Bell, E. A., & Thorwart, M. L. (1999). Water incorporated into a food but not served with a food decreases energy intake in lean women. *The American Journal of Clinical Nutrition, 70*, 448–455.

Rolls, B. J., Bell, E. A., & Waugh, B. A. (2000). Increasing the volume of a food by incorporating air affects satiety in men. *The American Journal of Clinical Nutrition, 72*, 361–368.

Rolls, B. J., Castellanos, V. H., Halford, J. C., Kilara, A., Panyam, D., Pelkman, C. L., et al. (1998). Volume of food consumed affects satiety in men. *The American Journal of Clinical Nutrition, 67*, 1170–1177.

Rolls, B. J., Ello-Martin, J. A., & Tohill, B. C. (2004). What can intervention studies tell us about the relationship between fruit and vegetable consumption and weight management? *Nutrition Reviews, 62*, 1–17.

Rolls, B. J., Roe, L. S., & Meengs, J. S. (2004). Salad and satiety: Energy density and portion size of a first-course salad affect energy intake at lunch. *Journal of the American Dietetic Association, 104*, 1570–1576.

Rolls, B. J., Roe, L. S., & Meengs, J. S. (2006). Reductions in portion size and energy density of foods are additive and lead to sustained decreases in energy intake. *The American Journal of Clinical Nutrition, 83*, 11–17.

Rolls, E. T. (2004). Smell, taste, texture, and temperature multimodal representations in the brain, and their relevance to the control of appetite. *Nutrition Reviews, 62*, S193–S204, discussion S224-141.

Ros, E. (2009). Nuts and novel biomarkers of cardiovascular disease. *The American Journal of Clinical Nutrition, 89*, 1649S–1656S.

Ros, E. (2010). Health benefits of nut consumption. *Nutrients, 2*, 652–682.

Rosell, M., Appleby, P., Spencer, E., & Key, T. (2006). Weight gain over 5 years in 21,966 meat-eating, fish-eating, vegetarian, and vegan men and women in EPIC-Oxford. *International Journal of Obesity, 30*, 1389–1396.

Rosen, L. A., Ostman, E. M., & Bjorck, I. M. (2011a). Effects of cereal breakfasts on postprandial glucose, appetite regulation and voluntary energy intake at a subsequent standardized lunch; focusing on rye products. *Nutrition Journal, 10*, 7.

Rosen, L. A., Ostman, E. M., & Bjorck, I. M. (2011b). Postprandial glycemia, insulinemia, and satiety responses in healthy subjects after whole grain rye bread made from different rye varieties. 2. *Journal of Agricultural and Food Chemistry, 59*, 12149–12154.

Rosen, L. A., Ostman, E. M., Shewry, P. R., Ward, J. L., Andersson, A. A., Piironen, V., et al. (2011). Postprandial glycemia, insulinemia, and satiety responses in healthy subjects after whole grain rye bread made from different rye varieties. 1. *Journal of Agricultural and Food Chemistry, 59*, 12139–12148.

Rosen, L. A., Silva, L. O., Andersson, U. K., Holm, C., Ostman, E. M., & Bjorck, I. M. (2009). Endosperm and whole grain rye breads are characterized by low post-prandial insulin response and a beneficial blood glucose profile. *Nutrition Journal, 8*, 42.

Sabate, J., Cordero-Macintyre, Z., Siapco, G., Torabian, S., & Haddad, E. (2005). Does regular walnut consumption lead to weight gain? *The British Journal of Nutrition, 94*, 859–864.

Sacks, F. M., Bray, G. A., Carey, V. J., Smith, S. R., Ryan, D. H., Anton, S. D., et al. (2009). Comparison of weight-loss diets with different compositions of fat, protein, and carbohydrates. *The New England Journal of Medicine, 360*, 859–873.

Sadiq Butt, M., Tahir-Nadeem, M., Khan, M. K., Shabir, R., & Butt, M. S. (2008). Oat: Unique among the cereals. *European Journal of Nutrition, 47*, 68–79.

Saltzman, E., Moriguti, J. C., Das, S. K., Corrales, A., Fuss, P., Greenberg, A. S., et al. (2001). Effects of a cereal rich in soluble fiber on body composition and dietary compliance during consumption of a hypocaloric diet. *Journal of the American College of Nutrition, 20*, 50–57.

Samra, R. A., & Anderson, G. H. (2007). Insoluble cereal fiber reduces appetite and short-term food intake and glycemic response to food consumed 75 min later by healthy men. *The American Journal of Clinical Nutrition, 86*, 972–979.

Samra, R. A., Wolever, T. M., & Anderson, G. H. (2007). Enhanced food intake regulatory responses after a glucose drink in hyperinsulinemic men. *International Journal of Obesity, 31*, 1222–1231.

Saquib, N., Natarajan, L., Rock, C. L., Flatt, S. W., Madlensky, L., Kealey, S., et al. (2008). The impact of a long-term reduction in dietary energy density on body weight within a randomized diet trial. *Nutrition and Cancer, 60*, 31–38.

Schroder, K. E. (2010). Effects of fruit consumption on body mass index and weight loss in a sample of overweight and obese dieters enrolled in a weight-loss intervention trial. *Nutrition, 26*, 727–734.

Schroeder, N., Gallaher, D. D., Arndt, E. A., & Marquart, L. (2009). Influence of whole grain barley, whole grain wheat, and refined rice-based foods on short-term satiety and energy intake. *Appetite, 53*, 363–369.

Schwarz, N. A., Rigby, B. R., La Bounty, P., Shelmadine, B., & Bowden, R. G. (2011). A review of weight control strategies and their effects on the regulation of hormonal balance. *Journal of Nutrition and Metabolism, 2011*, 237932.

Sclafani, A. (2004). Oral and postoral determinants of food reward. *Physiology and Behavior, 81*, 773–779.

Seimon, R. V., Lange, K., Little, T. J., Brennan, I. M., Pilichiewicz, A. N., Feltrin, K. L., et al. (2010). Pooled-data analysis identifies pyloric pressures and plasma cholecystokinin concentrations as major determinants of acute energy intake in healthy, lean men. *The American Journal of Clinical Nutrition, 92*, 61–68.

Sichieri, R. (2002). Dietary patterns and their associations with obesity in the Brazilian city of Rio de Janeiro. *Obesity Research, 10*, 42–48.

Simmons, A. L., Miller, C. K., Clinton, S. K., & Vodovotz, Y. (2011). A comparison of satiety, glycemic index, and insulinemic index of wheat-derived soft pretzels with or without soy. *Food and Function, 2*, 678–683.

Skendi, A., Biliaderis, C. G., Lazaridou, A., & Izydorczyk, M. S. (2002). Structure and rheological properties of water soluble Beta glucans from oat cultivars of *Avena sativa* and *Avena bysantina*. *Journal of Cereal Science, 38*, 15–31.

Slavin, J. L., & Green, H. (2007). Dietary fibre and satiety. *Nutrition Bulletin, 32*(Suppl. 1), 32–42.

Smeets, A. J., Soenen, S., Luscombe-Marsh, N. D., Ueland, O., & Westerterp-Plantenga, M. S. (2008). Energy expenditure, satiety, and plasma ghrelin, glucagon-like peptide 1, and peptide tyrosine-tyrosine concentrations following a single high-protein lunch. *The Journal of Nutrition, 138*, 698–702.

Smeets, A. J., & Westerterp-Plantenga, M. S. (2006). Satiety and substrate mobilization after oral fat stimulation. *The British Journal of Nutrition, 95*, 795–801.

Smeets, A. J., & Westerterp-Plantenga, M. S. (2009). The acute effects of a lunch containing capsaicin on energy and substrate utilisation, hormones, and satiety. *European Journal of Nutrition, 48*, 229–234.

Snitker, S., Fujishima, Y., Shen, H., Ott, S., Pi-Sunyer, X., Furuhata, Y., et al. (2009). Effects of novel capsinoid treatment on fatness and energy metabolism in humans: Possible pharmacogenetic implications. *The American Journal of Clinical Nutrition, 89*, 45–50.

Soenen, S., & Westerterp-Plantenga, M. S. (2007). No differences in satiety or energy intake after high-fructose corn syrup, sucrose, or milk preloads. *The American Journal of Clinical Nutrition, 86*, 1586–1594.

Song, S. W., Bae, Y. J., & Lee, D. T. (2010). Effects of caloric restriction with varying energy density and aerobic exercise on weight change and satiety in young female adults. *Nutrition Research and Practice, 4*, 414–420.

Song, W. O., Chun, O. K., Obayashi, S., Cho, S., & Chung, C. E. (2005). Is consumption of breakfast associated with body mass index in US adults? *Journal of the American Dietetic Association, 105*, 1373–1382.

St-Onge, M. P., & Jones, P. J. (2002). Physiological effects of medium-chain triglycerides: Potential agents in the prevention of obesity. *The Journal of Nutrition, 132*, 329–332.

Stewart, J. E., Feinle-Bisset, C., & Keast, R. S. (2011). Fatty acid detection during food consumption and digestion: Associations with ingestive behavior and obesity. *Progress in Lipid Research, 50*, 225–233.

Stock, M. J. (1999). Gluttony and thermogenesis revisited. *International Journal of Obesity and Related Metabolic Disorders, 23*, 1105–1117.

Strik, C. M., Lithander, F. E., McGill, A. T., MacGibbon, A. K., McArdle, B. H., & Poppitt, S. D. (2010). No evidence of differential effects of SFA, MUFA or PUFA on post-ingestive satiety and energy intake: A randomised trial of fatty acid saturation. *Nutrition Journal, 9*, 24.

Stubbs, R. J., Johnstone, A. M., O'Reilly, L. M., & Poppitt, S. D. (1998). Methodological issues relating to the measurement of food, energy and nutrient intake in human laboratory-based studies. *The Proceedings of the Nutrition Society, 57*, 357–372.

Stunkard, A. J., & Messick, S. (1985). The three-factor eating questionnaire to measure dietary restraint, disinhibition and hunger. *Journal of Psychosomatic Research, 29*, 71–83.

Summerbell, C. D., Douthwaite, W., Whittaker, V., Ells, L. J., Hillier, F., Smith, S., et al. (2009). The association between diet and physical activity and subsequent excess weight gain and obesity assessed at 5 years of age or older: A systematic review of the epidemiological evidence. *International Journal of Obesity, 33*(Suppl. 3), S1–S92.

Svetkey, L. P., Stevens, V. J., Brantley, P. J., Appel, L. J., Hollis, J. F., Loria, C. M., et al. (2008). Comparison of strategies for sustaining weight loss: The weight loss maintenance randomized controlled trial. *Journal of the American Medical Association, 299*, 1139–1148.

Thomas, D. E., Elliott, E. J., & Baur, L. (2007). Low glycaemic index or low glycaemic load diets for overweight and obesity. *Cochrane Database of Systematic Reviews, (3)* CD005105.

Thomas, D. M., Bouchard, C., Church, T., Slentz, C., Kraus, W. E., Redman, L. M., Martin, C. K., Silva, A. M., Vossen, M., Westerterp, K., & Heymsfield, S. B. (2012). Why do individuals not lose more weight from an exercise intervention at a defined dose? An energy balance analysis. *Obesity reviews, 13*, 835–847.

Thorne, M. J., Thompson, L. U., & Jenkins, D. J. (1983). Factors affecting starch digestibility and the glycemic response with special reference to legumes. *The American Journal of Clinical Nutrition, 38*, 481–488.

Thorsdottir, I., Tomasson, H., Gunnarsdottir, I., Gisladottir, E., Kiely, M., Parra, M. D., et al. (2007). Randomized trial of weight-loss-diets for young adults varying in fish and fish oil content. *International Journal of Obesity, 31*, 1560–1566.

Tohill, B. C., Seymour, J., Serdula, M., Kettel-Khan, L., & Rolls, B. J. (2004). What epidemiologic studies tell us about the relationship between fruit and vegetable consumption and body weight. *Nutrition Reviews, 62*, 365–374.

Tome, D., Schwarz, J., Darcel, N., & Fromentin, G. (2009). Protein, amino acids, vagus nerve signaling, and the brain. *The American Journal of Clinical Nutrition, 90*, 838S–843S.

Tordoff, M. G. (2001). Calcium: Taste, intake, and appetite. *Physiological Reviews, 81*, 1567–1597.

Tosh, S. M., Brummer, Y., Miller, S. S., Regand, A., Defelice, C., Duss, R., et al. (2010). Processing affects the physicochemical properties of beta-glucan in oat bran cereal. *Journal of Agricultural Food Chemistry, 58*, 7723–7730.

Traoret, C. J., Lokko, P., Cruz, A. C., Oliveira, C. G., Costa, N. M., Bressan, J., et al. (2008). Peanut digestion and energy balance. *International Journal of Obesity, 32*, 322–328.

Tsai, A. G., Williamson, D. F., & Glick, H. A. (2011). Direct medical cost of overweight and obesity in the USA: A quantitative systematic review. *Obesity Reviews*, *12*, 50–61.

Tsuchiya, A., Almiron-Roig, E., Lluch, A., Guyonnet, D., & Drewnowski, A. (2006). Higher satiety ratings following yogurt consumption relative to fruit drink or dairy fruit drink. *Journal of the American Dietetic Association*, *106*, 550–557.

Uhe, A. M., Collier, G. R., & O'Dea, K. (1992). A comparison of the effects of beef, chicken and fish protein on satiety and amino acid profiles in lean male subjects. *The Journal of Nutrition*, *122*, 467–472.

USDA-DHHS, (2010). *US Department of Agriculture and US Department of Health and Human Services. Dietary Guidelines for Americans 2010* (7th ed.). Washington, DC: US Government Printing Office. December 2010.

USDA, (2011). *Choosemyplate.gov, Food Groups.* http://www.choosemyplate.gov/food-groups Accessed on September 17, 2012.

USDA, (2012). *United States Department of Agriculture. Center for Nutrition Policy and Promotion. Report of the Dietary Guidelines Advisory Committee on the Dietary Guidelines for Americans 2010.* Accessed on September 17, 2012. http://www.cnpp.usda.gov/Publications/DietaryGuidelines/2010/DGAC/Report/D-5-Carbohydrates.pdf.

Van Citters, G. W., & Lin, H. C. (1999). The ileal brake: A fifteen-year progress report. *Current Gastroenterology Reports*, *1*, 404–409.

van Dam, R. M., & Seidell, J. C. (2007). Carbohydrate intake and obesity. *European Journal of Clinical Nutrition*, *61*(Suppl. 1), S75–S99.

Vander Wal, J. S., Gupta, A., Khosla, P., & Dhurandhar, N. V. (2008). Egg breakfast enhances weight loss. *International Journal of Obesity*, *32*, 1545–1551.

Vander Wal, J. S., Marth, J. M., Khosla, P., Jen, K. L., & Dhurandhar, N. V. (2005). Short-term effect of eggs on satiety in overweight and obese subjects. *Journal of the American College of Nutrition*, *24*, 510–515.

Vander Wal, J. S., Waller, S. M., Klurfeld, D. M., McBurney, M. I., Cho, S., Kapila, M., et al. (2006). Effect of a post-dinner snack and partial meal replacement program on weight loss. *International Journal of Food Sciences and Nutrition*, *57*, 97–106.

Velasquez, M. T., & Bhathena, S. J. (2007). Role of dietary soy protein in obesity. *International Journal of Medical Sciences*, *4*, 72–82.

Veldhorst, M. A., Nieuwenhuizen, A. G., Hochstenbach-Waelen, A., van Vught, A. J., Westerterp, K. R., Engelen, M. P., et al. (2009). Dose-dependent satiating effect of whey relative to casein or soy. *Physiology and Behavior*, *96*, 675–682.

Veldhorst, M. A., Nieuwenhuizen, A. G., Hochstenbach-Waelen, A., Westerterp, K. R., Engelen, M. P., Brummer, R. J., et al. (2009). Effects of high and normal soyprotein breakfasts on satiety and subsequent energy intake, including amino acid and 'satiety' hormone responses. *European Journal of Nutrition*, *48*, 92–100.

Veldhorst, M. A., Westerterp, K. R., & Westerterp-Plantenga, M. S. (2011). Gluconeogenesis and protein-induced satiety. *The British Journal of Nutrition*, *107*, 595–600.

Venn, B. J., & Green, T. J. (2007). Glycemic index and glycemic load: Measurement issues and their effect on diet-disease relationships. *European Journal of Clinical Nutrition*, *61*(Suppl. 1), S122–S131.

Venn, B. J., Perry, T., Green, T. J., Skeaff, C. M., Aitken, W., Moore, N. J., et al. (2010). The effect of increasing consumption of pulses and wholegrains in obese people: A randomized controlled trial. *Journal of the American College of Nutrition*, *29*, 365–372.

Vergnaud, A. C., Norat, T., Romaguera, D., Mouw, T., May, A. M., Romieu, I., et al. (2012). Fruit and vegetable consumption and prospective weight change in participants of the European prospective investigation into cancer and nutrition-physical activity, nutrition, alcohol, cessation of smoking, eating out of home, and obesity study. *The American Journal of Clinical Nutrition*, *95*, 184–193.

Vergnaud, A. C., Norat, T., Romaguera, D., Mouw, T., May, A. M., Travier, N., et al. (2010). Meat consumption and prospective weight change in participants of the EPIC-PANACEA study. *The American Journal of Clinical Nutrition, 92*, 398–407.

Verhoef, S. P., & Westerterp, K. R. (2011). No effects of Korean pine nut triacylglycerol on satiety and energy intake. *Nutrition and Metabolism, 8*, 79.

Vioque, J., Weinbrenner, T., Castello, A., Asensio, L., & Garcia de la Hera, M. (2008). Intake of fruits and vegetables in relation to 10-year weight gain among Spanish adults. *Obesity (Silver Spring), 16*, 664–670.

Viskaal-van Dongen, M., Kok, F. J., & de Graaf, C. (2010). Effects of snack consumption for 8 weeks on energy intake and body weight. *International Journal of Obesity, 34*, 319–326.

Vitaglione, P., Lumaga, R. B., Montagnese, C., Messia, M. C., Marconi, E., & Scalfi, L. (2010). Satiating effect of a barley beta-glucan-enriched snack. *Journal of the American College of Nutrition, 29*, 113–121.

Vitaglione, P., Lumaga, R. B., Stanzione, A., Scalfi, L., & Fogliano, V. (2009). beta-Glucan-enriched bread reduces energy intake and modifies plasma ghrelin and peptide YY concentrations in the short term. *Appetite, 53*, 338–344.

Wal, J. S., McBurney, M. I., Cho, S., & Dhurandhar, N. V. (2007). Ready-to-eat cereal products as meal replacements for weight loss. *International Journal of Food Sciences and Nutrition, 58*, 331–340.

Waller, S. M., Vander Wal, J. S., Klurfeld, D. M., McBurney, M. I., Cho, S., Bijlani, S., et al. (2004). Evening ready-to-eat cereal consumption contributes to weight management. *Journal of the American College of Nutrition, 23*, 316–321.

Wanders, A. J., van den Borne, J. J., de Graaf, C., Hulshof, T., Jonathan, M. C., Kristensen, M., et al. (2011). Effects of dietary fibre on subjective appetite, energy intake and body weight: A systematic review of randomized controlled trials. *Obesity Reviews, 12*, 724–739.

Wang, X., Li, Z., Liu, Y., Lv, X., & Yang, W. (2012). Effects of pistachios on body weight in Chinese subjects with metabolic syndrome. *Nutrition Journal, 11*, 20.

Wang, Y., & Beydoun, M. A. (2009). Meat consumption is associated with obesity and central obesity among US adults. *International Journal of Obesity, 33*, 621–628.

Wang, Y., Beydoun, M. A., Liang, L., Caballero, B., & Kumanyika, S. K. (2008). Will all Americans become overweight or obese? Estimating the progression and cost of the US obesity epidemic. *Obesity (Silver Spring), 16*, 2323–2330.

Warwick, Z. S., Hall, W. G., Pappas, T. N., & Schiffman, S. S. (1993). Taste and smell sensations enhance the satiating effect of both a high-carbohydrate and a high-fat meal in humans. *Physiology and Behavior, 53*, 553–563.

Weigle, D. S., Breen, P. A., Matthys, C. C., Callahan, H. S., Meeuws, K. E., Burden, V. R., et al. (2005). A high-protein diet induces sustained reductions in appetite, ad libitum caloric intake, and body weight despite compensatory changes in diurnal plasma leptin and ghrelin concentrations. *The American Journal of Clinical Nutrition, 82*, 41–48.

Welch, I., Saunders, K., & Read, N. W. (1985). Effect of ileal and intravenous infusions of fat emulsions on feeding and satiety in human volunteers. *Gastroenterology, 89*, 1293–1297.

Welch, I. M., Sepple, C. P., & Read, N. W. (1988). Comparisons of the effects on satiety and eating behaviour of infusion of lipid into the different regions of the small intestine. *Gut, 29*, 306–311.

Westerterp-Plantenga, M., Diepvens, K., Joosen, A. M., Berube-Parent, S., & Tremblay, A. (2006). Metabolic effects of spices, teas, and caffeine. *Physiology and Behavior, 89*, 85–91.

Westerterp-Plantenga, M. S., Nieuwenhuizen, A., Tome, D., Soenen, S., & Westerterp, K. R. (2009). Dietary protein, weight loss, and weight maintenance. *Annual Review of Nutrition, 29*, 21–41.

Westerterp-Plantenga, M. S., Rolland, V., Wilson, S. A., & Westerterp, K. R. (1999). Satiety related to 24 h diet-induced thermogenesis during high protein/carbohydrate vs high fat diets measured in a respiration chamber. *European Journal of Clinical Nutrition*, *53*, 495–502.

Westerterp-Plantenga, M. S., Smeets, A., & Lejeune, M. P. (2005). Sensory and gastrointestinal satiety effects of capsaicin on food intake. *International Journal of Obesity*, *29*, 682–688.

WHO, (2011). *Global Health Observatory Data Repository*. http://apps.who.int/ghodata/. Accessed on September 26, 2012.

Whole-Grain-Council, (2009). *Make half your grains whole conference*. http://www. wholegrainscouncil.org/files/3.AreWeThereYet.pdf. Accessed on September 18, 2012.

Wien, M. A., Sabate, J. M., Ikle, D. N., Cole, S. E., & Kandeel, F. R. (2003). Almonds vs complex carbohydrates in a weight reduction program. *International Journal of Obesity and Related Metabolic Disorders*, *27*, 1365–1372.

Wolever, T. M., Jenkins, D. J., Jenkins, A. L., & Josse, R. G. (1991). The glycemic index: Methodology and clinical implications. *The American Journal of Clinical Nutrition*, *54*, 846–854.

Wong, C. L., Mollard, R. C., Zafar, T. A., Luhovyy, B. L., & Anderson, G. H. (2009). Food intake and satiety following a serving of pulses in young men: Effect of processing, recipe, and pulse variety. *Journal of the American College of Nutrition*, *28*, 543–552.

Wood, P. J. (2004). Relationships between solution properties of cereal beta-glucans and physiological effects—A review. *Trends in Food Science and Technology*, *15*, 313–320.

Wood, P. J. (2007). Cereal beta-glucans in diet and health. *Journal of Cereal Science*, *46*, 230–238.

Yeomans, M. R., Blundell, J. E., & Leshem, M. (2004). Palatability: Response to nutritional need or need-free stimulation of appetite? *The British Journal of Nutrition*, *92*(Suppl. 1), S3–S14.

Yoshioka, M., Lim, K., Kikuzato, S., Kiyonaga, A., Tanaka, H., Shindo, M., et al. (1995). Effects of red-pepper diet on the energy metabolism in men. *Journal of Nutritional Science and Vitaminology*, *41*, 647–656.

Yoshioka, M., St-Pierre, S., Drapeau, V., Dionne, I., Doucet, E., Suzuki, M., et al. (1999). Effects of red pepper on appetite and energy intake. *The British Journal of Nutrition*, *82*, 115–123.

Yoshioka, M., St-Pierre, S., Suzuki, M., & Tremblay, A. (1998). Effects of red pepper added to high-fat and high-carbohydrate meals on energy metabolism and substrate utilization in Japanese women. *The British Journal of Nutrition*, *80*, 503–510.

Young, V. R., & Pellett, P. L. (1994). Plant proteins in relation to human protein and amino acid nutrition. *The American Journal of Clinical Nutrition*, *59*, 1203S–1212S.

Zemel, M. B. (2004). Role of calcium and dairy products in energy partitioning and weight management. *The American Journal of Clinical Nutrition*, *79*, 907S–912S.

CHAPTER FOUR

Biotransformation of Polyphenols for Improved Bioavailability and Processing Stability

Apoorva Gupta, Lalit D. Kagliwal, Rekha S. Singhal[1]
Food Engineering and Technology Department, Institute of Chemical Technology, Matunga, Mumbai, India
[1]Corresponding author: e-mail address: rsinghal7@rediffmail.com

Contents

Abstract

Research on the functions and effects of polyphenols has gained considerable momentum in recent times. This is attributed to their bioactivities, ranging from antioxidant to anticancer activities. But their potential is seldom fully realized since their solubility and stability is quite low and their bioavailability is hampered due to extensive metabolism in the body. Biotransformation of polyphenols using enzymes, whole cell microbes, or plant cell cultures may provide an effective solution by modifying their structure while maintaining their original bioactivity. Lipase, protease, cellulase, and transferases are commonly used enzymes, with lipase being the most popular for carrying out acylation reactions. Among the whole cell microbes, *Aspergillus*, *Bacillus*, and *Streptomyces* sp. are the most widely used, while *Eucalyptus perriniana* and *Capsicum frutescens* are the plant cell cultures used for the production of secondary metabolites. This chapter emphasizes the development of green solvents and identification of different sources/approaches to maximize polyphenol transformation for varied applications.

Advances in Food and Nutrition Research, Volume 69
ISSN 1043-4526
http://dx.doi.org/10.1016/B978-0-12-410540-9.00004-1

1. INTRODUCTION

Polyphenols belong to the class of secondary metabolites, which form an integral part of plant defense mechanism. As the name suggests, they have more than one phenolic hydroxyl group attached to one or more aromatic rings. They are believed to be the bioactive compounds responsible for antioxidant (Rice-Evans, Miller, & Paganga, 1997), antimicrobial (Mellou et al., 2005), anticancer (Fuggetta et al., 2006; Kunnumakkara et al., 2007; Schwarz & Roots, 2003), anti-inflammatory (Kanda et al., 1998), and antiulcer activities (Alarcon, Martin, Lacasa, & Motilva, 1994; Parmar & Parmar, 1998). They also exert a positive effect on neurodegenerative disorders such as Alzheimer's disease (Butterfield et al., 2002). Due to their structure, their most notable activity is antioxidant, that is, scavenging free radicals. The body is a constant source of free radicals generated during several physiological processes such as mitochondrial electron transport, which can lead to oxidative stress. The body combats this by producing superoxide dismutase, catalase, and glutathione transferase as intrinsic antioxidants. Polyphenols present in the diet supplement these systems. Found in numerous fruits, vegetables, legumes, and seeds such as soy and coffee, they are the most abundant micronutrients present in regular diet (Manach, Scalbert, Morand, Rémésy, & Jimenez, 2004). Their abundance in the diet together with their beneficial effects has contributed to the growing interest in and awareness of them. For instance, the relatively low incidence of coronary heart disease among the French, despite a diet rich in saturated fats, is attributed to the consumption of red wine, which is rich in polyphenols, namely, resveratrol.

In spite of a substantial daily intake of polyphenols, their benefits are not fully realized due to several issues that need to be addressed. Their bioavailability is greatly hampered as the polyphenols are metabolized to a large extent before reaching the target tissue. These metabolites, which are formed as a result of digestive or hepatic activity, differ from their native forms and may or may not be active. Also, their absorption from the intestine and elimination from the body depend highly on their structure (Manach et al., 2004). Polyphenols are generally found in plants in their glycosylated (a sugar attached to the polyphenol skeleton), aglycone (absence of sugar) or to a lesser extent, esterified forms. The glycosylated forms have poor solubility in lipophilic formulations and are soluble in polar solvents. Aglycones show poor solubility in water, while the esterified forms are very large and more lipid-soluble than water soluble. The form in which polyphenols are present in plants determines their behavior in different solvents. Thus, the barriers to bioavailability include solubility, permeability, metabolism, excretion, and target tissue uptake. Besides poor bioavailability, polyphenols exhibit low solubility and stability in commercial preparations, which restricts their use in many applications (Chebil, Humeau, Falcimaigne, Engasser, & Ghoul, 2006; Ishihara & Nakajima, 2003).

Biotransformation of polyphenols addresses these issues by altering the structure of polyphenols while maintaining their original bioactivity. The structure of a polyphenols can be altered by incorporation of a hydrophobic/hydrophilic group in their native structure. The resultant compound may have amphiphilic properties which will greatly expand its commercial application. Till date, researchers have attempted to modify polyphenols using chemical methods and enzymatic methods, as well as whole cell microbes and plant cells.

Additional advantages include increase in bioactivity, for instance, increase in radical scavenging or metal chelation activity (Mellou et al., 2005) by the formation of another polyphenol that is more effective (Brooks, Doyle, & O'Connor, 2006). Free polyunsaturated fatty acids (PUFAs) present in food are prone to oxidation and may pose health risks by creating free radicals. Esterification of poyphenols using these PUFAs can stabilize these easily oxidizable acids (Speranza & Macedo, 2012; Viskupičová, Danihelova, Ondrejovic, Liptaj, & Sturdik, 2010). Such enzymatic modification is also expected to reduce the astringency and bitterness that is present in polyphenols-rich foods (Degenhardt, Ullrich, Hofmann, & Stark, 2007). Such modifications also increase the diversity of polyphenols due to the formation of novel structures.

2. CLASSIFICATION OF POLYPHENOLS

Polyphenols are found abundantly in nature and in diverse forms. They can either be classified on the basis of their structure, source of origin, or physicochemical function. On the basis of structure, they can be subdivided into groups according to the number of phenolic rings and the structural elements that link these rings:

1. Phenolic acids
   a. Hydroxybenzoic acid derivatives (e.g., gallic acid)
   b. Hydroxycinnamic acid derivatives (e.g., caffeic acid, ferulic acid)
2. Flavonoids
   a. Isoflavones, neoflavonoids, and chalcones (e.g., genistein, diadzein, dalbergin, phloridzin)

   b. Flavones, flavonols, flavanones and flavanols (e.g., luteolin, quercetin, tangeretin, hesperetin, taxifolin)
   c. Flavanols and proanthocyanidins (e.g., catechin, epigallocatechin
   gallate (EGCG), condensed tannins)
   d. Anthocyanidins (e.g., malvidin, cyanidin)
3. Stilbenes (e.g., resveratrol)
4. Lignans (e.g., secoisolariciresinol)
5. Others (e.g., curcumin, hydroxytyrosol)

The chemical structures of polyphenols and their sources are given in Table 4.1.

The main dietary sources of polyphenols are fruits (such as apples, grapes, and berries), vegetables (such as onion and lettuce), and beverages (such as tea, coffee, and wine). One-third of our intake of polyphenols is dominated by phenolic acids and the remaining two-thirds by the largest subclass of flavonoids. These are present in food either in their free, glycosylated or esterified form, which dictates their passage, absorption, and bioavailability in the body.

3. BIOTRANSFORMATION OF POLYPHENOLS: STRATEGIES

3.1. Chemical methods

Chemical methods are less often used due to their nonselectivity and adverse environmental impact. Nevertheless, a number of researchers have used this method to synthesize polyphenols with novel structures. A higher degree of substitution in esterification reactions has been achieved using chemical methods as compared to enzymatic methods (Zhong & Shahidi, 2011). Chemical methods that have been investigated can be broadly classified as conjugation with proteins/amino acids, esterification, and polymerization. Table 4.2 summarizes chemical methods employed for the transformation of polyphenols.

3.1.1 Conjugation with proteins/amino acids

Kim et al. (2009) synthesized nine quercetin–amino acid conjugates and studied their pharmacokinetic properties. Amino acids were converted to their corresponding activated urethanes and the conjugates were obtained by alcoholysis with quercetin. They found that the conjugates possessing aspartic acid and glutamic acid showed 45.2- and 53.0-fold increase, respectively, in water solubility compared to quercetin. In a similar study (Koutsas et al., 2007), resveratrol conjugates with two different peptides—RGD

Table 4.1 Structural classification of polyphenols and their dietary sources

Polyphenol class	Dietary sources	Structure
Phenolic acids		
a. Hydroxybenzoic acid derivatives, e.g., gallic acid	Black tea, red wine	Gallic acid Caffeic acid Ferulic acid
b. Hydroxycinnamic acid derivatives, e.g., caffeic acid, ferulic acid	Coffee, wheat bran, blueberries	
Flavonoids		
a. Isoflavones, neoflavonoids, and chalcones, e.g., genistein, diadzein, dalbergin, phloridzin	Soyabean, soyamilk, apple	Genistein Diadzein Phloridzin
b. Flavones, flavonols, flavanones, and flavanols, e.g., luteolin, quercetin, tangeretin, hesperetin, taxifolin	Parsley, celery, mandarin, apple, onion, orange, grapefruit	Luteolin Tangeretin Hesperetin

Continued

Table 4.1 Structural classification of polyphenols and their dietary sources—cont'd

Polyphenol class	Dietary sources	Structure		
c. Flavanols and proanthocyanidins, e.g., catechin, epigallocatechin gallate, condensed tannins	Apricot, cherry, chocolate, red wine, green tea, grape seed	Quercetin	Taxifolin	Catechin
d. Anthocyanidins, e.g., malvidin, cyanidin	Orange, grapes, red wine, blackcurrant, blackberries	Epigallocatechin gallate	Malvidin	Cyanidin

Stilbenes

E.g. resveratrol	Grapes, berries, peanuts	trans-resveratrol

Lignans

E.g., secoisolariciresinol | Linseed

Secoisolariciresinol

Others

E.g., curcumin, hydroxytyrosol | Turmeric, olive oil, red and white wine, beer

Curcumin

Hydroxytyrosol

Table 4.2 Chemical methods employed for transformation of polyphenols

Polyphenol	Chemical reaction	Reaction outcome	References
Quercetin	Amino acid conjugates of quercetin synthesized (e.g., quercetin–glutamic acid conjugate)	Increases in water solubility and cell permeability as compared with parent molecule	Kim, Park, Yeo, Choo, and Chong (2009)
Resveratrol	Arg-Gly-Asp (RGD) and Lys-Gly-Asp (KGD) derivatives of resveratrol synthesized	– (increased health benefits anticipated)	Koutsas, Sarigiannis, Stavropoulos, and Liakopoulou-Kyriakides (2007)
EGCG (epigallocatechin gallate)	Esterification of EGCG with stearic acid, eicosapentaenoic acid, and docosahexanoic acid	Enhanced lipophilicity of derivatives with increased radical scavenging activity as compared to parent molecule	Zhong and Shahidi (2011)
Catechin	Polymerization of catechin	Increased superoxide scavenging activity as well as xanthine oxidase inhibitory activity as compared to monomeric catechin	Chung, Kurisawa, Kim, Uyama, and Kobayashi (2004)
Gallic acid and catechin	Gallic acid and catechin covalently bound to gelatin	Comparable antioxidant activity	Spizzirri et al. (2009)
Resveratrol	Addition of a glucosyl group to resveratrol via a succinate linker	Increase in water solubility	Biasutto et al. (2009)
Resveratrol	Synthesis of resveratrol coupled to cation triphenylphosphonium	Derivatives have good stability and are toxic only to fast-growing cancer cells	Biasutto et al. (2008)

(Arg-Gly-Asp) and KGD (Lys-Gly-Asp)—were synthesized. As these peptides are known to inhibit human platelet aggregation *in vitro*, their conjugates with resveratrol has been speculated to have synergistic activity.

Protein can be used to synthesize conjugates with polyphenols, which is expected to confer novel properties on the resulting molecule. In order to do so, gallic acid and catechin were covalently bound to gelatin separately in a grafting reaction (Spizzirri et al., 2009). The conjugate was found to have comparable antioxidant activity to the control polymer used.

3.1.2 Esterification
EGCG is the predominant catechin present in green tea and exhibits good solubility in hydrophilic systems, thus limiting its use in protecting lipophilic systems. Esterification of EGCG with different acyl donors such as stearic acid, eicosapentaenoic acid, and docosahexanoic acid showed an increase in partition coefficient of the resulting esters over the parent EGCG molecule. The antioxidant activity of the resultant esters was also found to be enhanced (Zhong & Shahidi, 2011). Similarly, conjugation of catechin with stearic acid strongly inhibited DNA polymerase, HL-60 cancer cell growth, and angiogenesis. It was also found to suppress human umbilical vein endothelial cell (HUVEC) tube formation on the reconstituted basement membrane, suggesting that it affected not only DNA polymerases but also signal transduction pathways in HUVECs (Matsubara et al., 2006).

Another chemical method was developed by Biasutto et al. (2009) where resveratrol was modified by the addition of a glucosyl group via a succinate linker. The addition of a glucosyl group aimed at increasing the water solubility, which influences its absorption. The product was 3,4′,5-tri (alpha-D-glucose-3-O-succinyl) resveratrol, which was found to be water soluble and nearly stable in an acidic environment, suggesting that it can survive the gastric stage with minimal destruction.

3.1.3 Polymerization
In recent times, biopolymers of flavonoids are also being synthesized. Poly (catechin) was prepared by condensation with acetaldehyde. The polymer exhibited superior superoxide scavenging activity as well as xanthine oxidase inhibitory activity as compared to monomeric catechin (Chung et al., 2004). Jin and Yoshioka (2005) synthesized poly-lauroyl-(+)-catechin in a similar manner. 3-Lauroyl-(+)-catechin (3-LC) and 3′,4′-lauroyl-(+)-catechin (3′,4′-LC) were the major products with 3-LC being a better antioxidant than catechin in the lipid phase.

3.1.4 Miscellaneous

Polyphenols are believed to possess anticancer properties (Hirose, Hoshiya, Akagi, Takalashi, & Hara, 1993). These properties may be realized either through their antioxidant or prooxidant actions. Resveratrol coupled to membrane-permeable lipophilic cation triphenylphosphonium was synthesized to target the resultant molecule to mitochondria. The cation was selected because it drives the accumulation of modified resveratrol in the mitochondrial matrix, where they are expected to fully realize their anticancer potential. The derivatives had good stability and were found to be cytotoxic only to fast-growing cancer cells (Biasutto et al., 2008).

Chemical methods are attractive in terms of cost and ease of operation but the hazards associated with the use of some toxic chemicals are very high, especially if the end product is intended for human use. The presence of numerous hydroxyl groups present in the polyphenol structure makes partial and/or selective modification rather difficult (Zhong & Shahidi, 2011). The hydroxyl groups present in EGCG at positions 5 and 7 on the A-ring are not so important in the radical scavenging activity (Nanjo et al., 1996). Thus, modifications made on these positions will not be expected to diminish the antioxidant activity. Another disadvantage of chemical methods is lesser control on the degree of substitution which may result in polyphenols with limited or no hydroxyl groups, leading to the loss of antioxidant or radical scavenging activity.

3.2. Enzymatic methods

The enzymatic strategy to transform polyphenols employs the enzymes isolated from microbes/plants and can also be carried out by immobilization of the enzyme. The use of enzymes offers several benefits such as wide substrate spectrum and ease of availability. Due to these reasons, enzymes have attracted a great deal of interest for structural modification and improvement of biochemical properties of polyphenols. As opposed to chemical methods, enzymes are more chemo-, regio-, and stereoselective. They operate under mild conditions of temperature and pressure and require no toxic solvents, thus making the process eco-friendly. They also make the purification simpler due to a higher specificity of reaction and few undesirable products. Microbial cells or enzymes can be immobilized and reused for many cycles. Various enzymes that have been tested for biocatalysis are lipase, transferase, isomerase, esterase, and protease, of which lipases are the most commonly used. Enzymatic biotransformation of polyphenols has been summarized in Table 4.3.

Table 4.3 Different enzymes employed for transformation of polyphenols

Enzyme	Source	Substrate	Product	Reaction outcome	References
Lipase	*Candida antarctica*	Rutin and naringin	Esters of oleic acid	Esters exhibited antiangiogenic and antitumor properties	Mellou, Loutrari, Stamatis, Roussos, and Kolisis (2006)
		Prunin	Ester of lauric acid	Increase in lipophilicity	Céliz and Daz (2011)
		Rutin	Esters of stearic acid, palmitic acid, linoleic acid, linolenic acid, oleic acid	Esters were more efficient antioxidants in oil matrix	Viskupičová et al. (2010)
		Isorhamnetin-3-O-glucoside	Esters of lauric and butyric acid	Increased inhibition of xanthine oxidase	Hadj Salem et al. (2011)
Tannase	*Aspergillus niger*	Catechin	Ester of gallic acid	Increase in antioxidant activity	Raab, Bel-Rhlid, Williamson, Hansen, and Chaillot (2007)
	Aspergillus ficuum	EGCG	Hydrolysis of EGCG to EGC and GA	Increase in antioxidant activity and cell viability	Lu and Chen (2008)
	Paecilomyces variotii	Green tea and yerba mate extracts	Hydrolysis of polyphenols	Increase in antioxidant activity	Macedo, Battestin, Ribeiro, and Macedo (2011)

Continued

Table 4.3 Different enzymes employed for transformation of polyphenols—cont'd

Enzyme	Source	Substrate	Product	Reaction outcome	References
Laccase	*Myceliopthora*	Rutin	Formation of rutin polymer	Enhanced radical scavenging activity	Kurisawa, Chung, Uyama, and Kobayashi (2003a)
	Unknown	Catechin	Gelatin–catechin conjugate formation	Amplified inhibition of LDL oxidation	Chung, Kurisawa, Uyama, et al. (2003)
Protease	*Bacillus subtilis*	Rutin	Ester formation with butyric acid	Ester induced higher frequency of micronuclei formation	Kodelia, Athanasiou, and Kolisis (1994)
		Troxerutin	Troxerutin acylated at the B-ethoxyl group	Higher conversion in presence of ultrasound treatment	Xiao, Yang, Mao, Zhao, and Lin (2011)
Transferase	Plant glycosyltransferases; *Eucalyptus perriniana*	Naringin and naringenin	Formation of beta glycosides	Improved solubility	Shimoda et al. (2010)
Peroxidase	Horseradish	Catechin	Formation of catechin conjugated with amine-substituted octahedral silsesquioxane	Enhanced superoxide anion scavenging activity, inhibition of xanthine oxidase activity	Ihara, Kurisawa, Chung, Uyama, and Kobayashi (2005)
Cellulase	*Trichoderma viride*	Catechin and EGCG	Formation of EGCG glucosides using dextrin	Glucosides showed improved heat stability, solubility, lower astringency than its aglycone	Noguchi et al. (2008)

Maltogenic amylase	*Bacillus stearothermophilus*	Neohesperidin dihydrochalcone (NHDC)	NHDC glycosylated with maltotriose	Maltosyl-NHDC was 700 times more soluble in water	Cho et al. (2000)
Tyrosinase	*Pseudomonas putida* F6	Tyrosol	Tyrosol converted to hydroxytyrosol	The hydroxytyrosol formed is a better antioxidant than tyrosol	Brooks et al. (2006)
Glucansucrase	*Leuconostoc mesenteroides* B-1299CB	EGCG	Glycosides of EGCG with sucrose	Glycosides exhibited greater water solubility than EGCG	Moon et al. (2006)
Hydrolase	*Aspergillus oryzae*	EGCG	EGCG hydrolyzed to EGC and GA	Increase in antioxidative capacity	Zhong, Shao, and Hong (2008)

3.2.1 Lipase

Lipases (triacylglycerol acylhydrolases, EC 3.1.1.3) belong to the class of serine hydrolases. They catalyze the hydrolysis of triacylglycerol to glycerol and fatty acids at the hydrophobic–hydrophilic interface. Under certain experimental conditions such as low water activity, they are capable of synthetic activity, that is, they form glycerides from fatty acids and glycerol (Sharma, Sharma, & Shukla, 2011). Industrial applications of lipase include bioremediation, biosurfactants, chiral separation, cheese ripening, paper, and pulp industries (Sharma et al., 2011). They are also being utilized for lipophilization of sugars, amino acids, and phenolic acids (Villeneuve, 2007). Microbial lipases have the largest application. *Candida antarctica* lipase B (CALB) is the most widely used biocatalyst in organic synthesis. The use of CALB offers several advantages. It can act on a wide range of substrates including rutin, naringin (Mellou et al., 2006), nonaqueous medium tolerance (Gayot, Santarelli, & Coulon, 2003), and thermal deactivation resistance (Ardhaoui et al., 2004; Kontogianni, Skouridou, Sereti, Stamatis, & Kolisis, 2003; Stevenson, Wibisono, Jensen, Stanley, & Cooney, 2006).

Polyphenols can be present in different forms—glycosylated, aglycone, or esterified. CALB is capable of acylating only glycosylated polyphenols. If aglycones are present, lipase from *Pseudomonas cepacia* can be used. When lipase from *P. cepacia* was employed for acylating catechin, an aglycone, the acylation took place on C5 and C7 hydroxyl groups (Lambusta, Nicolosi, Patti, & Piatelli, 1993).

Rutin and naringin have been acylated using oleic acid as the acyl donor, using an immobilized lipase B from *Candida antarctica* in acetone at 50 °C. The synthesized esters were found to be more efficient in reducing the release of Vascular Endothelial Growth Factor (VEGF) from K562 cells than their parent molecule indicating antiangiogenic and antitumor properties (Mellou et al., 2006). Kontogianni et al. (2003) acylated naringin and rutin with medium-chain fatty acids and they found water activity of 0.11 or lower to favor the conversion which followed the Michelis–Menten kinetics.

Céliz and Daz (2011) described the preparation of alkyl esters of prunin, a flavanone glucoside derivative of naringin in organic media using two different immobilized lipases: CALB and *Rhizomucor meihei*. Vinyl laurate was the acyl donor and the synthesized derivatives retained their antioxidant activity with increased lipophilicity. Viskupičová et al. (2010) esterified rutin with different fatty acids in 2-methylbutan-2-ol. Esters of rutin with stearate, palmitate, oleate, linoleate, and linolenate were found to be the most efficient antioxidant agents in an oil matrix.

Water and methanol extracts/fractions of *Nitraria retusa* have also been investigated for their flavonoid content (Hadj Salem et al., 2011). The enzymatic acylation of isorhamnetin-3-O-glucoside with ethyl laurate and ethyl butyrate increased its capacity to inhibit xanthine oxidase and its antiproliferative activity but reduced its radical scavenging activity.

Acylation reactions are very sensitive in the presence of water since its presence can drive the reaction backward. Hence, molecular sieves are often employed in such reactions. It can increase the reaction rate and yield. In the synthesis of rutin stearate, they had a positive impact on both the reaction rate and yield (Duan, Du, Yao, Li, & Wu, 2006).

Recently, enzyme catalysis using lipase in room-temperature ionic liquids (RTILs) has attracted a lot of attention as "green solvents." Unlike conventional organic solvents, they possess no vapor pressure, can dissolve a variety of molecules, and form two-phase systems with many solvents (Kragl, Eckstein, & Kaftzik, 2002). Rutin and esculin were esterified with palmitic and oleic acids in 14 different ionic liquids. RTILs containing TF2N−, PF6−, and BF4− anions were most successful as reaction media and exhibited an increase in yield (Lue, Guo, & Xu, 2010).

Yang, Guo, and Xu (2012) examined fatty alcohols instead of fatty acids for lipase catalyzed esterification reactions. Esters of phenolic acids, ferulic, and dihydrocaffeic acids were synthesized using C4–C18 straight-chain fatty alcohols. They found the conversion to be dependent on the number of carbon atoms in fatty alcohol. Maximum conversion of 95% was achieved when hexanol was used. Similarly, catechins were also modified using *n*-butanol. 3-O-Acylcatechins were synthesized in the presence of immobilized *Mucor meihei* lipase, Lipozyme IM (Patti, Piattelli, & Nicolosi, 2000).

3.2.2 Tannase

Tannase (tannin acylhydrolase, EC 3.1.1.20) catalyzes the hydrolysis of bonds present in hydrolyzable tannins and gallic acid esters (Lekha & Lonsane, 1997). It is an extracellular, inducible enzyme produced by bacteria, fungi, and yeast in the presence of tannic acid and has shown to have considerable synthetic potential. It is extensively used in tanning, wine, beer, and as an additive in food detanification. For tannase production by *Paecilomyces variotii*, temperature range of 40–65 °C and pH range of 4.5–6.5 have been reported as optimal (Battestin & Macedo, 2007).

A tannase from *Aspergillus niger* was used for esterification of catechin using gallic acid in ionic liquids, a process known as galloylation (Raab et al., 2007). An increase in esterification was observed with immobilization of tannase. Further increase in esterification up to 6% was observed with increasing concentration of gallic acid. Water content greater than 20% shifted the thermodynamic equilibrium toward hydrolysis of the ester products, while lower water content lead to inactivation of the enzyme.

Interestingly, some polyphenols are also believed to be capable of prooxidant activities. EGCG, a green tea polyphenol, is known to possess very good antioxidant activity. However, in the presence of metal ion complexes of Cu^{2+} and Fe^{3+}, it causes DNA damage (Furukawa, Oikawa, Murata, Hiraku, & Kawanishi, 2003). To combat this, tannase was employed to reduce the relative percentage of EGCG in green tea extract. When one mole of EGCG is hydrolyzed, one mole each of epigallocatechin (EGC) and gallic acid are produced. Due to this, the resultant concentration of catechins in the hydrolysate increases (increase in EGC and gallic acid (GA)), which is further expected to increase the antioxidant activity. The effect of EGCG reduction was demonstrated by evaluating the viability of RAW 264.7 cells using both enzyme treated and nontreated green tea extract in the presence of Cu^{2+}. Viability of cells was higher than the control used (Lu & Chen, 2008). In a similar study, Macedo et al. (2011) evaluated the antioxidant properties of green tea and yerba mate extracts before and after enzyme treatment catalyzed by the *P. variotii* tannase. The antioxidant power of the enzyme treated green tea and yerba mate extracts increased by 55% and 43%, respectively, as compared to that of untreated extracts. This increase in activity is due to the increase in resultant concentration of catechins.

3.2.3 Laccase

Laccase (EC 1.10.3.2) belongs to the blue multicopper oxidases and its functions include degradation of polymers, ring cleavage of aromatic compounds, and cross-linking of monomers (Shekher, Sehgal, Kamthania, & Kumar, 2011). This ability to cross-link monomers has been utilized for transforming polyphenol skeletons. Catechin was enzymatically conjugated to one of the several available lysine residues of gelatin (Chung, Kurisawa, Uyama, & Kobayashi, 2003). Gelatin was used with the view to produce a highly functional biopolymer as it possesses some desirable properties such as biodegradability, low level of immunogenicity, and cytotoxicity. The conjugate was found to have an amplified inhibition against Low-density lipoprotein (LDL) oxidation. However, superoxide anion radical scavenging activity of the conjugate was slightly inferior to that of unmodified catechin. Similarly, the synthesis of poly(allylamine)–catechin conjugate has also been investigated (Chung, Kurisawa, Tachibana, Uyama, & Kobayashi, 2003).

Polymers of flavonoids have also been synthesized using laccase. Rutin was polymerized using *Myceliophthora* laccase to produce a water-soluble polymer, which exhibited a greater amplified radical scavenging activity than rutin monomer against superoxide anion (Kurisawa et al., 2003a). This could be attributed to the creation of a higher concentration of phenolic moieties in the molecules. Similar effects were seen when catechin was polymerized using laccase (Kurisawa, Chung, Uyama, & Kobayashi, 2003b).

3.2.4 Esterases and proteases

Esterases hydrolyze an ester into its corresponding acid and alcohol. Carboxylesterases include two groups of enzymes, namely, the nonspecific esterases and lipases. Both of them possess hydrolytic capability but in the absence of water, they are able to reverse the reaction and synthesize esters. The difference lies in specificity toward the substrate. While esterases act on short acyl chain esters, lipases act on long-chain triacylglycerols (Chahiniana & Sarda, 2009). Since esterases are capable of incorporating only a short hydrophobic chain to the polyphenols, they are not attractive for the purpose of enzymatic transformation.

Even though the physiological role of proteases is hydrolysis of water-soluble proteins, they have been used as catalysts in nonaqueous solvents. This replaces their hydrolytic activity with synthetic activity such as peptide synthesis, and esterification of sugars. Subtilisin is the most commonly used alkaline protease. It was shown to introduce a butyryl moiety to rutin, a flavonoid diglucoside and esterification occurred exclusively on the sugar moiety (Kodelia & Kolisis, 1992). The cytotoxic effect of such an ester was explored by Kodelia et al. (1994). They found no difference between rutin and its ester in inducing sister chromatid exchange. However, the frequency of micronuclei formation increased drastically from 3.5% for rutin to 8% for rutin ester at a similar dose. This was speculated to be due to increased penetration of ester through the living cell membrane, owing to its amphiphilic character.

Comparative studies of enzymatic acylation of troxerutin by the alkaline protease from *Bacillus subtilis* were carried out by Xiao et al. (2011). Acylation was done both under ultrasound as well as shaking in nonaqueous media. Divinyl dicarboxylates of varying carbon-chain lengths were used as acyl donors. Troxerutin was found to be acylated at the B-ethoxyl group in both reaction conditions. Troxerutin conversion increased under ultrasonic treatment.

3.2.5 Other enzymes

3.2.5.1 Transferases

Transferases are enzymes that transfer a functional group from one compound (donor) to another (acceptor). In biotransformation, the subclass of glycosyltransferases are the most frequently used. They are a large family of enzymes involved in biosynthesis of oligosaccharides, polysaccharides, and glycol conjugates. Found in both prokaryotes and eukaryotes, they transfer a sugar moiety by forming a glycosidic bond (Breton, Šnajdrová, Jeanneau, Koča, & Imberty, 2006). The transfer of sugar to the polyphenol skeleton enhances the water solubility and stability of aglycones and flavonoids (Xiao, Cao, Wang, Zhao, & Wei, 2009). Glycosylation occurs naturally in plants and serves various purposes such as reducing toxicity of xenobiotics, increasing solubility and stability of endogenous aglycones.

3.2.5.2 Peroxidase

Green tea catechin was enzymatically conjugated with amine-substituted octahedral silsesquioxane using horseradish peroxidase (Ihara et al., 2005). The conjugate exhibited enhanced superoxide anion scavenging activity as well as more than eightfold inhibition in xanthine oxidase activity.

3.2.5.3 Cellulase

Cellulase obtained from *Trichoderma viride* has been used to *trans*-glucosylate (+)-catechin and (−)-epigallocatechin gallate using dextrin as a glucosyl donor (Noguchi et al., 2008). Some of the EGCG derivatives, such as EGCG 5-O-α-D-glucopyranoside, showed improved heat stability and solubility, lower astringency, and astringent stimulation than its aglycone, suggesting that EGCG glucosides is functionally superior to EGCG as food additives.

Enzymatic transformations are affected by the type, origin, and concentration of enzymes, nature of polyphenol, acyl donor, and optimal conditions of temperature, water activity, and pH. Immobilization of enzymes offers several benefits such as reusability, higher productivity, specificity, and simple isolation, making them an attractive choice for biotransformation.

3.3. Factors affecting enzyme-mediated transformation of polyphenols

Various factors affecting enzyme-mediated transformation of polyphenols are the type and origin of the biocatalyst, the structure and concentration of the substrates (acyl donor, acyl acceptor, and their ratio), the reaction media, and the reaction temperature.

3.3.1 Acyl donor
Higher conversion yields of esculin and rutin (>70%) were obtained with aliphatic acids having high carbon-chain length (>12) (Ardhaoui et al., 2004). CALB was found to exhibit a high activity toward short- and medium-chain length acyl donors and was less active toward long-chain substrates (Salem et al., 2010).

3.3.2 Acyl acceptor
The initial rate and conversion yield of flavonoid increases with an increase in its concentration. However, the concentration of flavonoid is limited by its solubility in the reaction medium (Chebil et al., 2006). But when solubilization is not a limiting factor, the initial conversion rate reaches a plateau for concentrations higher than 120 mM of naringin (Kontogianni et al., 2003).

3.3.3 Solvent
The reaction medium should be such that it allows appropriate solubility of polar acyl acceptor (flavonoid glycoside) and nonpolar acyl donor as well as the highest possible enzyme activity. Most frequently used solvents are 2-methylbutan-2-ol and acetone because of their low toxicity.

3.3.4 Water activity
Lipases catalyze hydrolytic activity in the presence of water, so very little water is required for the synthetic activity of lipase. The optimum water content for esterification in organic solvents is 0.2–3%. Ardhaoui et al. (2004) observed the best lipase activity at 200 ppm of water.

3.3.5 Temperature
It affects the viscosity of reaction medium, enzyme stability, and substrate and product solubility. *C. antarctica* is a thermostable enzyme, thus high catalytic activities are observed with an increase in temperature. The time needed for biotransformation of phloridzin to phloridzin cinnamate depends on the temperature applied. An increase in temperature from 60 to 100 °C decreases the time of biotransformation from 15 to 2 h (Enaud, Humeau, Piffaut, & Girardin, 2004).

3.4. Microbes
Microbial transformation is an effective tool, besides enzymes, for the structural modification of bioactive natural and synthetic compounds, including flavonoids. Microbial production offers several advantages such as rapid

growth, ease of cultivation, convenient genetic manipulations, and high-level production. Moreover, it increases product selectivity, reduces use of toxic chemicals, and conserves energy. Biotransformations of polyphenols by many microorganisms including different species of *Aspergillus* and *Bacillus* have been examined and investigated. A summary of microbes involved in the biotransformation of polyphenols is given in Table 4.4.

3.4.1 Fungi
3.4.1.1 Aspergillus sp.
Commercial lipases and esterases from *Aspergillus* sp. are effective in biotransformation of polyphenols. Biotransformation of the isoflavones diadzein, 7,4-dimethoxyisoflavone, and 7,4,3-diacetoxyisoflavone was carried out using *A. niger* as a biocatalyst (Miyazawa et al., 2004). Baqueiro-Peña et al. (2010) reported the formation of vanillic acid and 4-vinylguaiacol (starting material for the production of vanillin) from ferulic acid by the action of both wild and a diploid strain of *A. niger*. The conversion of ferulic acid was 64% for the diploid strain, while for the wild strain it was 36%.

3.4.1.2 Paecilomyces variotii
Besides the use of *A. niger*, a filamentous fungus *P. variotii* MTCC 6581 was also analyzed for its capability to transform ferulic acid into vanillate derivatives (Ghosh et al., 2006). When cultures of *P. variotii* were grown on minimal medium containing ferulic acid (10 mM) as sole carbon source, formation of vanillic acid was observed. *In vitro* conversion of ferulic acid to vanillic acid was also demonstrated with a cell-free extract.

3.4.2 Bacteria
3.4.2.1 Bacillus sp.
Bacillus sp. is used in the biotransformation of polyphenols for their safety, rapid growth, and ease of scale-up for mass cultures. An intestinal bacterium in chicken hydrolyses the flavonoid diglycosides present in the feed. Three anaerobic *Lactobacillus*-like strains designated as MF-01, MF-02, and MF-03 from the cecum of chicken could convert rutin and hesperidin into their respective aglyconic forms (Iqbal & Zhu, 2009).

Generally, human intestinal enzymes do not have β-glycosidase activity, which is necessary for hydrolysis of the bond between the sugar moiety and the bioactive compound in flavonoid glycosides. Thus, the uptake is affected and so the health benefits are not fully realized. Hence, Di Gioia et al. (2010) investigated 15 strains of the commonly found lactic acid bacteria for the

Table 4.4 Microorganisms involved in biotransformation of polyphenols

Microorganism	Substrate	Product	Outcome	References
Bacteria				
Lactobacillus sp.	Rutin and hesperidin	Conversion into aglyconic forms	Increased availability of rutin and hesperidin	Iqbal and Zhu (2009)
L. buchneri and *L. acidophilus*	Wheat flavonoid glycosides	Conversion into aglyconic forms	Increased uptake of flavonoids	Di Gioia, Bregola, Aloisio, Marotti, and Dinelli (2010)
Streptomyces sp.	Naringenin, quercetin, and some stilbenoids (resveratrol, rhapontigenin, deoxyrhapontigenin)	Glucuronidated products formed	–	Marvalin and Azerad (2011)
Streptomyces sp.	Phloretin and chrysin	Hydroxylated phloretin and chrysin	–	Kim (2010)
Pseudomonas aeruginosa	Tyrosol	Formation of hydroxytyrosol	Hydroxytyrosol is a more potent antioxidant	Allouche, Damak, Ellouz, and Sayadi (2004)
Fungi				
Aspergillus niger	Diadzein,7, 4-dimethoxyisoflavone, 7,4,3-diacetoxyisoflavone	Diadzein was not transformed while the others gave several metabolites	–	Miyazawa, Ando, Okuno, and Araki (2004)

Continued

Table 4.4 Microorganisms involved in biotransformation of polyphenols—cont'd

Microorganism	Substrate	Product	Outcome	References
A. niger	Ferulic acid	Formation of vanillic acid and 4-vinyl guaiacol	Products are starting materials for production of commercially important compound, vanillin	Baqueiro-Peña et al. (2010)
Paecilomyces variotii	Ferulic acid	Formation of vanillic acid	Starting material for vanillin	Ghosh, Sachan, and Mitra (2006)
A. niger and *Penicillium chermesinum*	Flavanone	Hydroxylated flavanones formed	—	Kostrzewa-Susłow, Dmochowska-Gładysz, Białońska, and Ciunik (2008)

presence of β-glycosidase enzyme activity. They found strains of *L. buchneri* and *L. acidophilus* to be efficient agents for biotransformation of the wheat flavonoid glycosides and hence potential candidates for functional food development. Similarly, *B. subtilis* natto NTU-18 in black soymilk could effectively hydrolyze the glycosides from isoflavone (Kuo, Cheng, Wu, Huang, & Lee, 2006).

3.4.2.2 *Streptomyces* sp.

Human detoxification mechanisms include conjugation and hence most of the flavonoids and stilbenes circulate in the blood as glucuronides or sulfates. Thus, their intrinsic activity needs to be elucidated and potential roles defined. For this, Marvalin and Azerad (2011) performed a one-step biotransformation of naringenin, quercetin, and some stilbenoids (resveratrol, rhapontigenin, deoxyrhapontigenin) by incubation with a *Streptomyces* sp. M52104 with the aim of polyphenol glucuronidation. *Streptomyces* sp. M52104 resulted in the formation of regioselective glucuronidated products. Kim (2010) used *Streptomyces* sp. for carrying out the biotransformation of phloretin and chrysin. Hydroxylated products were formed with 12% conversion yield and characterized by using HPLC and GC-MS.

Recently, metabolic engineering approaches have resulted in transgenic microorganisms expressing different transferases, thereby eliminating the need for performing biotransformation as the secreted flavonoid itself is conjugated. Prenyltransferase genes from *Streptomyces* were transferred and overexpressed in a model legume, *Lotus japonicas*. Further, two genes from the *Sophora flavescens* plant were also overexpressed in *L. japonicas*. It was found to produce prenylated polyphenols when supplemented with exogenous polyphenols (Sugiyama et al., 2011). Similarly, the biotransformation of naringenin was carried out by using transgenic yeast containing the prenyltransferase gene from *S. flavescens* (Sasaki, Tsurumaru, & Yazaki, 2009).

3.5. Plant cell cultures

As discussed earlier, glycosyltransferases present in plants serve various purposes such as increasing the solubility and stability of aglycones. Plant cell cultures are utilized for the same purpose. They transform exogenously fed polyphenols by conjugating a sugar residue with them. Glycosylation is a characteristic biotransformation reaction of plant cells. Two types of glycosyl conjugations are present. In the first type, the esterification reaction occurs between a carboxylic acid and sugar moiety and in the second type, ether is formed (glycosylation) from the hydroxyl group and sugar moiety (Ishihara, Hamada, Hirata, & Nakajima, 2003).

The advantages of employing plant cell cultures are several, such as
- Cells can be grown in the laboratory to an almost unlimited quantity that can give large amounts of the desired product.
- Cultured cells are homogenous, with negligible variation in the quality of products (Ishihara et al., 2003).

However, due to engineering challenges in scale-up, economic feasibility has still not been established. One particular challenge is the formation of aggregates by plant cells, which influence culture productivity since aggregates are not adequately exposed to the light needed to induce flavonoid biosynthesis (Hanagata et al., 1993). Plant cells involved in bioconversion of polyphenols are discussed in Table 4.5.

Table 4.5 Plant cells involved in biotransformation of polyphenols

Plant	Substrate	Product	Reaction outcome	References
Eucalyptus perriniana	Naringin and naringenin	β-Glycosides were formed	Improved aqueous solubility	Shimoda et al. (2010)
	Hesperitin	Glycosides of hesperitin	Improved aqueous solubility	Shimoda, Hamada, and Hamada (2008)
	Diadzein	Glycosylated diadzein formed	–	Shimoda, Sato, Kobayashi, Hamada, and Hamada (2008)
Citrus aurantium L.	Naringenin and hesperitin	Glucosylated products formed	Increased aqueous solubility	Lewinsohn, Berman, Mazur, and Gressel (1989)
Capsicum frutescens	Protocatechuic aldehyde and caffeic acid	Bioconversion into vanillin and capsaicin	Vanillin and capsaicin are commercially important compounds	Rao and Ravishankar (2000)
	Caffeic acid and veratraldehyde	Biotransformation to vanillin and other related metabolites	Vanillin is a commercially important compound	Suresh and Ravishankar (2005)

Till date, plant cells from *Eucalyptus perriniana* (Shimoda et al., 2010), *Citrus paradisi* (Lewinsohn, Berman, Mazur, & Gressel, 1986), *Capsicum frutescens* (Rao & Ravishankar, 2000), and *Catharanthus roseus* (Shimoda, Yamane, Hirakawa, Ohta, & Hirata, 2002) have been utilized for the biotransformation of exogenously fed polyphenols.

3.5.1 Eucalyptus perriniana

Shimoda et al. (2010) carried out the biotransformation of naringin and naringenin by using *E. perriniana* cell cultures for β-glycosides production. Hesperitin has very low solubility in aqueous media and oral absorption is also low. When *Ipomea batatas* and *E. perriniana* cell cultures were used for biotransformation of hesperitin, regioselective glycosylated products were formed, which can overcome the barriers of hesperitin (Shimoda, Hamada, & Hamada, 2008).

The sequential glycosylation of soybean isoflavone, daidzein, with cultured suspension cells of *E. perriniana* and cyclodextrin glucanotransferase, was studied by Shimoda, Sato, Kobayashi, Hamada, and Hamada (2008). Diadzein was found to be glycosylated by *E. perriniana* cells. Glycosylation of vanillin and 8-nordihydrocapsaicin was also carried out by cultured *E. perriniana* cells (Sato et al., 2012).

3.5.2 Capsicum frutescens

Rao and Ravishankar (2000) used both freely suspended cells and immobilized cells of *C. frutescens*. These cells were incubated with phenyl-propanoid intermediates—protocatechuic aldehyde and caffeic acid—which were biotransformed by the cultures into vanillin and capsaicin. Similarly, Suresh and Ravishankar (2005) used *C. frutescens* root culture biotransformation of caffeic acid and veratraldehyde to vanillin and other related metabolites. Methyl jasmonate was found to act as an elicitor for the bioconversion of caffeic acid.

3.5.3 Citrus paradisi

Other plant cell cultures used for biotransformation are *C. paradisi* (grapefruit) cells (Lewinsohn et al., 1986). In a study by Lewinsohn et al. (1989), the ability of undifferentiated citrus cultures to biotransform flavanones was carried out. They found that only one undifferentiated sour orange (*Citrus aurantium* L.) ovule-derived cell culture was able to glucosylate exogenous naringenin and also rhamnosylated the product further, synthesizing narirutin. Other ovule-derived cell lines, one each from sour orange and a hybrid, were able to glucosylate exogenous naringenin and hesperitin. Glucosylation occurred only in the presence of exogenously supplied polyphenols.

4. IMPACT OF BIOTRANSFORMATION

4.1. Metabolite production

Metabolite production is a very important consequence of the biotransformation of polyphenols. Several reports are available on the accumulation of metabolites such as vanillin and capsaicin, which are widely used in the food industry. Vanillin is the primary component of vanilla bean extract. Synthetic vanillin is sometimes used as a flavoring agent in foods, beverages, and pharmaceuticals. Capsaicin is the active component of chilli peppers and is an irritant for mammals, including humans. It produces a sensation of burning in any tissue with which it comes into contact. It is commonly used in food products to give them added spice or "heat." The production of metabolites is majorly observed from plant cells while fungi and algae are also used.

C. frutescens has been extensively used for carrying out biotransformations. When precursors like caffeic acid and veratraldehyde are externally fed to root cultures, vanillic acid, and other related metabolites were formed. Methyl jasmonate treatment increased the production of vanillin (Suresh & Ravishankar, 2005). Similarly, freely suspended and immobilized cell cultures of C. frutescens were fed externally with caffeic acid and protocatechuic aldehyde for biotransformation into vanillin and capsaicin (Rao & Ravishankar, 2000).

A white rot fungus Trametes sp. converted ferulic acid into coniferyl alcohol which was further degraded to vanillic acid (Nishida & Fukuzumi, 1978). Degradation of ferulic acid by another white rot fungus, Pycnoporus cinnabarinus was shown to catabolize ferulic acid (Falconnier et al., 1994). 4-Coumaric acid has also been biotransformed into caffeic acid using P. cinnabarinus cultures (Estrada Alvarado et al., 2003).

Another fungus, Sporotrichum thermophile, metabolized ferulic acid by propenoic chain degradation. 4-Vinylguaiacol formation was also observed. Guaiacol was also observed in the media due to nonoxidative decarboxylation of vanillic acid. Under optimum conditions, conversion was high and the molar yield was 35% (Topakas, Kalogeris, Kekos, Macris, & Christakopoulos, 2003).

A filamentous fungus P. variotii MTCC 6581 also exhibited similar reactions where ferulic acid was transformed into vanillate derivatives. When minimal medium was used to culture P. variotii using only ferulic acid as the main carbon source, vanillic acid was found to be accumulated. Similar results were observed when cell-free extracts were used (Ghosh et al., 2006).

Vanilla flavor metabolites were also reported as a result of biotransformation carried out by microalga, *Haematococcus pluvalis*, in free and immobilized forms (Tripathi, Ramachandra Rao, & Ravishankar, 2002). Phenylpropanoid intermediates such as ferulic acid and *p*-coumaric acid were fed, which were biotransformed to vanillin, vanillic acid, and vanillyl alcohol.

4.2. Increased bioactivity

Biotransformation may result in a polyphenol having better antioxidant activity than the parent compound. Researchers have used different approaches, including the use of enzymes and microorganisms, for studying the potential of polyphenols.

Brooks et al. (2006) used cell extracts of *Pseudomonas putida* F6 expressing tyrosinase activity to convert tyrosol into hydroxytyrosol. Hydroxytyrosol is a more potent antioxidant than tyrosol. They studied various parameters including addition of ascorbic acid, washing and reuse of immobilized cell extracts, and concentration of tyrosol for maximizing the production of hydroxytyrosol.

The use of the enzyme tannase to increase the antioxidant activity of tea polyphenols has also been reported (Lu & Chen, 2008). Green tea was used since it contains more polyphenols than fermented black tea. They reported that the tannase-treated green tea possessed higher antioxidant capacity (superoxide anion, hydrogen peroxide, and radical scavenging) than the untreated tea. Further, the viability of RAW 264.7 cells incubated with tannase-treated green tea was also higher compared to those incubated with untreated tea.

Esterification of polyphenols with saturated or unsaturated fatty acid makes them lipophilic, which helps their application in oil-based foods. Zhong, Ma, and Shahidi (2012) extracted the water-soluble green tea polyphenol, EGCG, and esterified it with different fatty acids. Esterification with long-chain PUFAs resulted in increased peroxyl radical scavenging activity. The EGCG esters even showed remarkably higher antiviral activities than the EGCG. These derivatives have also shown inhibitory activity against α-glucosidase, suggesting their potential in anti-HIV treatment.

Acylation of flavonoids can also lead to potentiation of their antimicrobial and antioxidant properties (Mellou et al., 2005). They have shown introduction of acyl group in glucosylated flavonoids to increase the antioxidant activity against LDL and serum models *in vitro*. Also, acylated disaccharidic flavonoid increases its antimicrobial properties against the Gram-positive microorganisms—*Staphylococcus aureus* and *Bacillus cereus*.

4.3. Improved bioavailability

Bioavailability of polyphenols is of great concern since very little is known about their effects *in vivo*. The body has a detoxification mechanism of its own for carrying out the metabolism of xenobiotics. Polyphenols, considered foreign compounds by the body, undergo phase I and phase II reactions that result in the formation of compounds which may or may not be bioactive. If they are not absorbed by the body and eliminated soon, no beneficial effect could be observed. The biological properties of polyphenols depend on their bioavailability. Indirect evidence of their absorption through the gut barrier is the increase in the antioxidant capacity of the plasma after the consumption of polyphenol-rich foods (Duthie et al., 1998; Young et al., 1999). More direct evidence can be obtained by measuring their concentrations in plasma and urine after the ingestion of either pure compounds or foodstuffs with a known content of polyphenols (Tian, Giusti, Stoner, & Schwartz, 2006). It is hypothesized that if polyphenols can be lipophilized, their ability to penetrate the lipid-containing cell membrane will be enhanced. Thus, a lot of work is being aimed at increasing the lipophilicity of polyphenols but *in vivo* studies are still lacking. Kodelia et al. (1994) demonstrated that synthesized lipophilic polyphenol (rutin ester) was able to penetrate the cell membrane more effectively. They found that the induced frequency of micronuclei formation in treated cells was much higher for rutin ester (8%) than for rutin (3.5%) for similar doses, which may be due to easier penetration across the cell membrane.

Plant polyphenols can be clinically exploited when their bioavailability is not hampered. If precursors of polyphenols that resist the metabolism occurring in phase II can be synthesized, the bioavailability may be enhanced. Biasutto, Marotta, Marchi, Zoratti, and Paradisi (2007) studied the stability and solubility of quercetin ester derivatives. Transport of these derivatives across supported tight monolayers of MDCK-1, MDCK-2, and Caco-2 cells was examined in order to study the transepithelial absorption. They found that only a few of the quercetin esters crossed the monolayers after undergoing partial deacylation and no phase II conjugation was observed, which was favorable.

4.4. Solubility and stability in lipophilic preparations

Céliz and Daz (2011) described the preparation of alkyl esters of prunin, a flavanone glucoside derivative of naringin in organic media using immobilized lipases with vinyl laurate as the donor. The modified esters

Table 4.6 Impact of biotransformation on polyphenols

Impact	Example	References
Metabolite production	*Capsicum frutescens* cells were externally fed with protocatechuic aldehyde and caffeic acid which led to their conversion into vanillin and capsaicin	Rao and Ravishankar (2000)
Stability and solubility in lipophilic matrices	Alkyl esters of prunin exhibited increased solubility in hydrophobic medium 1-octanol	Céliz and Daz (2011)
Increase in bioavailability	Induced frequency of micronuclei formation in rutin ester than for rutin	Kodelia et al. (1994)
Increased bioactivity	Cell extracts of *Pseudomonas putida* converted tyrosol into hydroxytyrosol which is a more potent antioxidant than tyrosol	Brooks et al. (2006)

retained radical scavenging ability and their solubility increased in hydrophobic medium 1-octanol.

Stevenson et al. (2006) performed direct acylation of flavonoid glycosides with palmitic, cinnamic, and phenylpropionic acids and its hydroxylated derivatives, catalyzed by immobilized CALB. They found that acylated derivatives had increased lipophilicity and solubility in reaction solvent.

Food preservation is necessary in lipophilic matrices since they are highly susceptible to lipid peroxidation, which causes qualitative decay. The addition of lipophilic antioxidant polyphenols may result in better storage of these products. Thus, lipases are used to esterify polyphenols. Viskupičová et al. (2010) esterified rutin with fatty acids in 2-methylbutan-2-ol. This resulted in the formation of esters of rutin with stearate, palmitate, oleate, linoleate, and linolenate with better antioxidants than rutin in an oil matrix. A summary of biotransformation of polyphenols and its impact on properties is given in Table 4.6.

5. CONCLUSION

Transformation of polyphenols can be carried out by using chemicals, enzymes, microorganisms, and plant cell culture. Biotransformation of polyphenols has attracted a lot of scientific and industrial interest. Change in the structure of polyphenols results in changes in their physicochemical as well as biological properties. Such changes produce polyphenols with increased lipophilicity, bioavailability, enhanced stability, solubility, and antioxidant activity, which greatly broaden the scope of utilization of polyphenols. However, the success of the modification depends on the method used for transformation of polyphenols. Chemical methods are attractive in terms of cost and ease of operation but the hazards involved due to some toxic chemicals are very high, if the end product is intended for human use. Transformation using enzymes, microbes, and even plant cells is a green choice since it involves mild, nonhazardous conditions and does not generate toxic wastes. Much progress has been made in this area with lipase emerging as the most widely used enzyme for esterifying polyphenols. *Streptomyces* sp. and *Aspergillus* sp. are the most widely used microorganisms and *E. perriniana* and *C. frutescens* are the most widely used plant cells. At present, researchers are focusing on the development of efficient solvent systems, and the identification of different sources/approaches for maximizing the production of polyphenol derivatives. However, an evaluation of bioactivity, bioavailability, and safety of such derivatives must follow and should form an integral part of the study.

REFERENCES

Alarcon, D. L. L. C., Martin, M. J., Lacasa, C., & Motilva, V. (1994). Antiulcerogenic activity of flavonoids and gastric protection. *Journal of Ethnopharmacology, 42*, 161–170.
Allouche, N., Damak, M., Ellouz, R., & Sayadi, S. (2004). Use of whole cells of Pseudomonas aeruginosa for synthesis of the antioxidant hydroxytyrosol via conversion of tyrosol. *Applied and Environmental Microbiology, 70*(4), 2105–2109.
Ardhaoui, M., Falcimaigne, A., Ognier, S., Engasser, J., Moussou, P., Pauly, G., et al. (2004). Effect of acyl donor chain length and substitutions pattern on the enzymatic acylation of flavonoids. *Journal of Biotechnology, 110*, 265–272.
Baqueiro-Peña, I., Rodríguez-Serrano, G., González-Zamora, E., Augur, C., Loera, O., & Saucedo-Castañeda, G. (2010). Biotransformation of ferulic acid to 4-vinylguaiacol by a wild and a diploid strain of *Aspergillus niger*. *Bioresource Technology, 101*(12), 4721–4724.
Battestin, V., & Macedo, G. A. (2007). Tannase production by *Paecilomyces variotii*. *Bioresource Technology, 98*(9), 1832–1837.
Biasutto, L., Marotta, E., Bradaschia, A., Fallica, M., Mattarei, A., Garbisa, S., et al. (2009). Soluble polyphenols: Synthesis and bioavailability of 3,4′,5-tri(alpha-D-glucose-3-O-succinyl) resveratrol. *Bioorganic and Medicinal Chemistry Letters, 19*(23), 6721–6724.
Biasutto, L., Marotta, E., Marchi, U. D., Zoratti, M., & Paradisi, C. (2007). Ester based precursors to increase the bioavailability of Quercetin. *Journal of Medicinal Chemistry, 50*, 241–253.
Biasutto, L., Mattarei, A., Marotta, E., Bradaschia, A., Sassi, N., Garbisa, S., et al. (2008). Development of mitochondria-targeted derivatives of resveratrol. *Bioorganic and Medicinal Chemistry Letters, 18*(20), 5594–5597.
Breton, C., Šnajdrová, L., Jeanneau, C., Koča, J., & Imberty, A. (2006). Structures and mechanisms of glycosyltransferases. *Glycobiology, 16*(2), 29R–37R.
Brooks, S. J., Doyle, E. M., & O'Connor, K. E. (2006). Tyrosol to hydroxytyrosol biotransformation by immobilised cell extracts of *Pseudomonas putida* F6. *Enzyme and Microbial Technology, 39*(2), 191–196.

Butterfield, D. A., Castegna, A., Pocernich, C. B., Drake, J., Scapagninib, G., & Calabresec, V. (2002). Nutritional approaches to combat oxidative stress in Alzheimer's disease. *Journal of Nutritional Biochemistry, 13*, 444–461.

Céliz, G., & Daz, M. (2011). Biocatalytic preparation of alkyl esters of citrus flavanone glucoside prunin in organic media. *Process Biochemistry, 46*(1), 94–100.

Chahiniana, H., & Sarda, L. (2009). Distinction between esterases and lipases: Comparative biochemical properties of sequence-related carboxylesterases. *Protein and Peptide Letters, 16*(10), 1149–1161.

Chebil, L., Humeau, C., Falcimaigne, A., Engasser, J., & Ghoul, M. (2006). Enzymatic acylation of flavonoids. *Process Biochemistry, 41*, 2237–2251.

Cho, J. S., Yoo, S. S., Cheong, T. K., Kim, M. J., Kim, Y., & Park, K. H. (2000). Transglycosylation of neohesperidin dihydrochalcone by *Bacillus stearothermophilus* maltogenic amylase. *Journal of Agricultural and Food Chemistry, 48*(2), 152–154.

Chung, J. E., Kurisawa, M., Kim, Y. J., Uyama, H., & Kobayashi, S. (2004). Amplification of antioxidant activity of catechin by polycondensation with acetaldehyde. *Biomacromolecules, 5*(1), 113–118.

Chung, J. E., Kurisawa, M., Tachibana, Y., Uyama, H., & Kobayashi, S. (2003). Enzymatic synthesis and antioxidant property of poly (allylamine)-catechin conjugate. *Chemistry Letters, 7*, 620–621.

Chung, J. E., Kurisawa, M., Uyama, H., & Kobayashi, S. (2003). Enzymatic synthesis and antioxidant property of gelatin-catechin conjugates. *Biotechnology Letters, 25*(23), 1993–1997.

Degenhardt, A., Ullrich, F., Hofmann, T., & Stark, T. (2007). Flavonoid sugar addition products, method for manufacture and use thereof. US Patent 20070269570.

Di Gioia, D., Bregola, V., Aloisio, I., Marotti, I., & Dinelli, G. (2010). Biotransformation of common bean and wheat flavonoid glycosides by lactic acid bacteria. *Journal of Biotechnology, 150*, 340.

Duan, Y., Du, Z., Yao, Y., Li, R., & Wu, D. (2006). Effect of molecular sieves on lipase-catalyzed esterification of rutin with stearic acid. *Journal of Agricultural and Food Chemistry, 54*(17), 6219–6225.

Duthie, G. G., Pedersen, M. W., Gardner, P. T., Morrice, P. C., Jenkinson, A. M., McPhail, D. B., et al. (1998). The effect of whisky and wine consumption on total phenol content and antioxidant capacity of plasma from healthy volunteers. *European Journal of Clinical Nutrition, 52*(10), 733–736.

Enaud, E., Humeau, C., Piffaut, B., & Girardin, M. (2004). Enzymatic synthesis of new aromatic esters of phloridzin. *Journal of Molecular Catalysis B: Enzymatic, 27*, 1–6.

Estrada Alvarado, I., Navarro, D., Record, E., Asther, M., Asther, M., & Lesage-Meessen, L. (2003). Fungal biotransformation of 4-coumaric acid into caffeic acid by *Pycnoporus cinnabarinus*: An alternative for producing a strong natural antioxidant. *World Journal of Microbiology and Biotechnology, 19*, 157–160.

Falconnier, B., Lapierre, C., Lesage-Meessen, L., Yonnet, G., Brunerie, P., Colonna-Ceccaldi, B., et al. (1994). Vanillin as a product of ferulic acid biotransformation by the white-rot fungus *Pycnoporus cinnabarinus* I-937: Identification of metabolic pathways. *Journal of Biotechnology, 37*(2), 123–132.

Fuggetta, M. P., Lanzilli, G., Tricarico, M., Cottarelli, A., Falchetti, R., Ravagnan, G., et al. (2006). Effect of resveratrol on proliferation and telomerase activity of human colon cancer cells in vitro. *Journal of Experimental and Clinical Cancer Research, 25*(2), 189–193.

Furukawa, A., Oikawa, S., Murata, M., Hiraku, Y., & Kawanishi, S. (2003). (−)-Epigallocatechin gallate causes oxidative damage to isolated and cellular DNA. *Biochemical Pharmacology, 66*(9), 1769–1778.

Gayot, S., Santarelli, X., & Coulon, D. (2003). Modification of flavonoid using lipase in nonconventional media: Effect of the water content. *Journal of Biotechnology, 101*, 29–36.

Ghosh, S., Sachan, A., & Mitra, A. (2006). Formation of vanillic acid from ferulic acid by *Paecilomyces variotii* MTCC. *Current Science*, *90*(6), 825–829.

Hadj Salem, J., Chevalot, I., Harscoat-Schiavo, C., Paris, C., Fick, M., & Humeau, C. (2011). Biological activities of flavonoids from *Nitraria retusa* (Forssk.) Asch. and their acylated derivatives. *Food Chemistry*, *124*(2), 486–494.

Hanagata, N., Ito, A., Uehara, H., Asari, F., Takeuchi, T., & Karube, I. (1993). Behavior of cell aggregate of *Carthamus tinctorius* L. cultured cells and correlation with red pigment formation. *Journal of Biotechnology*, *30*(3), 259–269.

Hirose, M., Hoshiya, T., Akagi, T., Takalashi, S., & Hara, Y. (1993). Effect of green tea catechin in a rat multi-organ carcinogenesis model. *Carcinogenesis*, *14*, 1549–1553.

Ihara, N., Kurisawa, M., Chung, J. E., Uyama, H., & Kobayashi, S. (2005). Enzymatic synthesis of a catechin conjugate of polyhedral oligomeric silsesquioxane and evaluation of its antioxidant activity. *Applied Microbiology and Biotechnology*, *66*(4), 430–433.

Iqbal, M. F., & Zhu, W. Y. (2009). Characterization of newly isolated *Lactobacillus delbrueckii*-like strain MF-07 isolated from chicken and its role in isoflavone biotransformation. *FEMS Microbiology Letters*, *291*(2), 180–187.

Ishihara, K., Hamada, H., Hirata, T., & Nakajima, N. (2003). Biotransformation using plant cultured cells. *Journal of Molecular Catalysis B: Enzymatic*, *23*(2), 145–170.

Ishihara, K., & Nakajima, N. (2003). Structural aspects of acylated plant pigments: Stabilization of flavonoid glucosides and interpretation of their functions. *Journal of Molecular Catalysis B: Enzymatic*, *23*, 411–417.

Jin, J., & Yoshioka, H. (2005). Synthesis of lipophilic poly-lauroyl-(+)-catechins and radical scavenging activity. *Bioscience, Biotechnology, and Biochemistry*, *69*(3), 440–447.

Kanda, T., Akiyama, H., Yanagida, A., Tanabe, M., Goda, Y., Toyoda, M., et al. (1998). Inhibitory effects of apple polyphenol on induced histamine release from RBL-2H3 cells and rat mast cells. *Bioscience, Biotechnology, and Biochemistry*, *62*(7), 1284–1289.

Kim, J. (2010). Biotransformation of flavonoid compounds; Phloretin and chrysin, using *Streptomyces* species. *Journal of Biotechnology*, *150*, 347.

Kim, M. K., Park, K. S., Yeo, W. S., Choo, H., & Chong, Y. (2009). In vitro solubility, stability and permeability of novel quercetin–amino acid conjugates. *Bioorganic and Medicinal Chemistry*, *17*(3), 1164–1171.

Kodelia, G., Athanasiou, K., & Kolisis, F. N. (1994). Enzymatic synthesis of butyryl-rutin ester in organic solvents and its cytogenetic effects in mammalian cells in culture. *Applied Biochemistry and Biotechnology*, *44*(3), 205–212.

Kodelia, G., & Kolisis, F. N. (1992). Studies on the reaction catalyzed by protease for the acylation of flavonoids in organic solvents. *Annals of the New York Academy of Sciences*, *672*(1), 451–457.

Kontogianni, A., Skouridou, V., Sereti, V., Stamatis, H., & Kolisis, F. N. (2003). Lipase-catalyzed esterification of rutin and naringin with fatty acids of medium carbon chain. *Journal of Molecular Catalysis B: Enzymatic*, *21*(1), 59–62.

Kostrzewa-Susłow, E., Dmochowska-Gładysz, J., Białońska, A., & Ciunik, Z. (2008). Microbial transformations of flavanone by *Aspergillus niger* and *Penicillium chermesinum* cultures. *Journal of Molecular Catalysis B: Enzymatic*, *52*, 34–39.

Koutsas, C., Sarigiannis, Y., Stavropoulos, G., & Liakopoulou-Kyriakides, M. (2007). Conjugation of resveratrol with RGD and KGD derivatives. *Protein and Peptide Letters*, *14*(10), 1014–1020.

Kragl, U., Eckstein, M., & Kaftzik, N. (2002). Enzyme catalysis in ionic liquids. *Current Opinion in Biotechnology*, *13*(6), 565–571.

Kunnumakkara, A. B., Guha, S., Krishnan, S., Diagaradjane, P., Gelovani, J., & Aggarwal, B. B. (2007). Curcumin potentiates antitumor activity of gemcitabine in an orthotopic model of pancreatic cancer through suppression of proliferation, angiogenesis, and inhibition of nuclear factor-κB-regulated gene products. *Cancer Research*, *67*(8), 3853–3861.

Kuo, L. C., Cheng, W. Y., Wu, R. Y., Huang, C. J., & Lee, K. T. (2006). Hydrolysis of black soybean isoflavone glycosides by *Bacillus subtilis natto*. *Applied Microbiology and Biotechnology*, *73*(2), 314–320.

Kurisawa, M., Chung, J. E., Uyama, H., & Kobayashi, S. (2003a). Enzymatic synthesis and antioxidant properties of poly (rutin). *Biomacromolecules*, *4*, 1394–1399.

Kurisawa, M., Chung, J. E., Uyama, H., & Kobayashi, S. (2003b). Laccase-catalyzed synthesis and antioxidant property of poly (catechin). *Macromolecular Bioscience*, *3*(12), 758–764.

Lambusta, D., Nicolosi, G., Patti, A., & Piatelli, M. (1993). Enzyme-mediated region-protection-deprotection of hydroxyl groups in (+)-catechin. *Synthesis*, *11*, 1155–1158.

Lekha, P. K., & Lonsane, B. K. (1997). Production and application of tannin acyl hydrolase: State of the art. *Advances in Applied Microbiology*, *44*, 215–260.

Lewinsohn, E., Berman, E., Mazur, Y., & Gressel, J. (1986). Glucosylation of exogenous flavanones by grapefruit (*Citrus paradisi*) cell cultures. *Phytochemistry*, *25*(11), 2531–2535.

Lewinsohn, E., Berman, E., Mazur, Y., & Gressel, J. (1989). Glucosylation and (1–6) rhamnosylation of exogenous flavanones by undifferentiated citrus cell cultures. *Plant Science*, *61*, 23–28.

Lu, M. J., & Chen, C. (2008). Enzymatic modification by tannase increases the antioxidant activity of green tea. *Food Research International*, *41*(2), 130–137.

Lue, B. M., Guo, Z., & Xu, X. (2010). Effect of room temperature ionic liquid structure on the enzymatic acylation of flavonoids. *Process Biochemistry*, *45*(8), 1375–1382.

Macedo, J. A., Battestin, V., Ribeiro, M. L., & Macedo, G. A. (2011). Increasing the antioxidant power of tea extracts by biotransformation of polyphenols. *Food Chemistry*, *126*(2), 491–497.

Manach, C., Scalbert, A., Morand, C., Rémésy, C., & Jimenez, L. (2004). Polyphenols: Food sources and bioavailability. *American Journal of Clinical Nutrition*, *79*, 727–747.

Marvalin, C., & Azerad, R. (2011). Microbial glucuronidation of polyphenols. *Journal of Molecular Catalysis B: Enzymatic*, *73*(1), 43–52.

Matsubara, K., Saito, A., Tanaka, A., Nakajima, N., Akagi, R., Mori, M., et al. (2006). Catechin conjugated with fatty acid inhibits DNA polymerase and angiogenesis. *DNA and Cell Biology*, *25*(2), 95–103.

Mellou, F., Lazari, D., Skaltsa, H., Tselepis, A., Kolisis, F., & Stamatis, H. (2005). Biocatalytic preparation of acylated derivatives of flavonoid glycosides enhances their antioxidant and antimicrobial activity. *Journal of Biotechnology*, *116*, 295–304.

Mellou, F., Loutrari, H., Stamatis, H., Roussos, C., & Kolisis, F. N. (2006). Enzymatic esterification of flavonoids with unsaturated fatty acids: Effect of the novel esters on vascular endothelial growth factor release from K562 cells. *Process Biochemistry*, *41*, 2029–2034.

Miyazawa, M., Ando, H., Okuno, Y., & Araki, H. (2004). Biotransformation of isoflavones by *Aspergillus niger*, as biocatalyst. *Journal of Molecular Catalysis B: Enzymatic*, *27*, 91–95.

Moon, Y. H., Lee, J. H., Ahn, J. S., Nam, S. H., Oh, D. K., Park, D. H., et al. (2006). Synthesis, structure analyses, and characterization of novel epigallocatechin gallate (EGCG) glycosides using the glucansucrase from *Leuconostoc mesenteroides* B-1299CB. *Journal of Agricultural and Food Chemistry*, *54*(4), 1230–1237.

Nanjo, F., Goto, K., Seto, R., Suzuki, M., Sakai, M., & Hara, Y. (1996). Scavenging effects of tea catechins and their derivatives on 1,1-diphenyl-2-picrylhydrazyl radical. *Free Radical Biology and Medicine*, *21*, 895–902.

Nishida, A., & Fukuzumi, T. (1978). Formation of coniferyl alcohol from ferulic acid by white rot fungus *Trametes*. *Phytochemistry*, *17*(3), 417–419.

Noguchi, A., Inohara-Ochiai, M., Ishibashi, N., Fukami, H., Nakayama, T., & Nakao, M. (2008). A novel glucosylation enzyme: Molecular cloning, expression, and characterization of *Trichoderma viride* JCM22452 α-amylase and enzymatic synthesis of some flavonoid monoglucosides and oligoglucosides. *Journal of Agricultural and Food Chemistry*, *56*(24), 12016–12024.

Parmar, N. S., & Parmar, S. (1998). Anti-ulcer potential of flavonoids. *Indian Journal of Physiology and Pharmacology*, *42*(3), 343–351.

Patti, A., Piattelli, M., & Nicolosi, G. (2000). Use of *Mucor miehei* lipase in the preparation of long chain 3-O-acylcatechins. *Journal of Molecular Catalysis B: Enzymatic*, *10*(6), 577–582.

Raab, T., Bel-Rhlid, R., Williamson, G., Hansen, C. E., & Chaillot, D. (2007). Enzymatic galloylation of catechins in room temperature ionic liquids. *Journal of Molecular Catalysis B: Enzymatic*, *44*(2), 60–65.

Rao, S. R., & Ravishankar, G. A. (2000). Biotransformation of protocatechuic aldehyde and caffeic acid to vanillin and capsaicin in freely suspended and immobilized cell cultures of *Capsicum frutescens*. *Journal of Biotechnology*, *76*(2), 137–146.

Rice-Evans, C., Miller, N., & Paganga, G. (1997). Antioxidant properties of phenolic compounds. *Trends in Plant Science*, *2*(4), 152–159.

Salem, J. H., Humeau, C., Chevalot, I., Harscoat-Schiavo, C., Vanderesse, R., Blanchard, F., et al. (2010). Effect of acyl donor chain length on isoquercitrin acylation and biological activities of corresponding esters. *Process Biochemistry*, *45*(3), 382–389.

Sasaki, K., Tsurumaru, Y., & Yazaki, K. (2009). Prenylation of flavonoids by biotransformation of yeast expressing plant membrane-bound prenyltransferase SfN8DT-1. *Bioscience, Biotechnology, and Biochemistry*, *73*(3), 759–761.

Sato, D., Eshita, Y., Katsuragi, H., Hamada, H., Shimoda, K., & Kubota, N. (2012). Glycosylation of vanillin and 8-nordihydrocapsaicin by cultured *Eucalyptus perriniana* cells. *Molecules*, *17*(5), 5013–5020.

Schwarz, D., & Roots, I. (2003). In vitro assessment of inhibition by natural polyphenols of metabolic activation of procarcinogens by human CYP1A1. *Biochemical and Biophysical Research Communications*, *303*(3), 902–907.

Sharma, D., Sharma, B., & Shukla, A. K. (2011). Biotechnological approach of microbial lipase: A review. *Biotechnology*, *10*(1), 23–40.

Shekher, R., Sehgal, S., Kamthania, M., & Kumar, A. (2011). Laccase: Microbial sources, production, purification, and potential biotechnological applications. *Enzyme Research*, Article ID 217861, 11 pages. http://dx.doi.org/10.4061/2011/217861.

Shimoda, K., Hamada, H., & Hamada, H. (2008). Glycosylation of hesperetin by plant cell cultures. *Phytochemistry*, *69*, 1135–1140.

Shimoda, K., Kubota, N., Taniuchi, K., Sato, D., Nakajima, N., Hamada, H., et al. (2010). Biotransformation of naringin & naringenin by cultured *Eucalyptus perriniana* cells. *Phytochemistry*, *71*(2), 201–205.

Shimoda, K., Sato, N., Kobayashi, T., Hamada, H., & Hamada, H. (2008). Glycosylation of daidzein by the *Eucalyptus* cell cultures. *Phytochemistry*, *69*(12), 2303–2306.

Shimoda, K., Yamane, S. Y., Hirakawa, H., Ohta, S., & Hirata, T. (2002). Biotransformation of phenolic compounds by the cultured cells of *Catharanthus roseus*. *Journal of Molecular Catalysis B: Enzymatic*, *16*(5), 275–281.

Speranza, P., & Macedo, G. A. (2012). Lipase-mediated production of specific lipids with improved biological and physicochemical properties. *Process Biochemistry*, *47*(12), 1699–1706.

Spizzirri, U. G., Iemma, F., Puoci, F., Cirillo, G., Curcio, M., Parisi, O. I., et al. (2009). Synthesis of antioxidant polymers by grafting of gallic acid and catechin on gelatin. *Biomacromolecules*, *10*(7), 1923–1930.

Stevenson, D., Wibisono, R., Jensen, D., Stanley, R., & Cooney, J. (2006). Direct acylation of flavonoid glycosides with phenolic acids catalysed by *Candida antarctica* lipase B (Novozym 435(R)). *Enzyme and Microbial Technology*, *39*, 1236–1241.

Sugiyama, A., Linley, P. J., Sasaki, K., Kumano, T., Yamamoto, H., Shitan, N., et al. (2011). Metabolic engineering for the production of prenylated polyphenols in transgenic legume plants using bacterial and plant prenyltransferases. *Metabolic Engineering, 13*(6), 629–637.

Suresh, B., & Ravishankar, G. A. (2005). Methyl jasmonate modulated biotransformation of phenylpropanoids to vanillin related metabolites using *Capsicum frutescens* root cultures. *Plant Physiology and Biochemistry, 43*(2), 125–131.

Tian, Q., Giusti, M. M., Stoner, G. D., & Schwartz, S. J. (2006). Urinary excretion of black raspberry (*Rubus occidentalis*) anthocyanins and their metabolites. *Journal of Agricultural and Food Chemistry, 54*(4), 1467–1472.

Topakas, E., Kalogeris, E., Kekos, D., Macris, B. J., & Christakopoulos, P. (2003). Bioconversion of ferulic acid into vanillic acid by the thermophilic fungus *Sporotrichum* thermophile. *LWT—Food Science and Technology, 36*(6), 561–565.

Tripathi, U., Ramachandra Rao, S., & Ravishankar, G. A. (2002). Biotransformation of phenylpropanoid compounds to vanilla flavor metabolites in cultures of *Haematococcus pluvialis*. *Process Biochemistry, 38*(3), 419–426.

Villeneuve, P. (2007). Lipases in lipophilization reactions. *Biotechnology Advances, 25*(6), 515–536.

Viskupičová, J., Danihelova, M., Ondrejovic, M., Liptaj, T., & Sturdik, E. (2010). Lipophilic rutin derivatives for antioxidant protection of oil-based foods. *Food Chemistry, 123*, 45–50.

Xiao, J., Cao, H., Wang, Y., Zhao, J., & Wei, X. (2009). Glycosylation of dietary flavonoids decreases the affinities for plasma protein. *Journal of Agricultural and Food Chemistry, 57* (15), 6642–6648.

Xiao, Y., Yang, L., Mao, P., Zhao, Z., & Lin, X. (2011). Ultrasound-promoted enzymatic synthesis of troxerutin esters in nonaqueous solvents. *Ultrasonics Sonochemistry, 18*(1), 303–309.

Yang, Z., Guo, Z., & Xu, X. (2012). Enzymatic lipophilisation of phenolic acids through esterification with fatty alcohols in organic solvents. *Food Chemistry, 132*(3), 1311–1315.

Young, J. F., Nielsen, S. E., Haraldsdóttir, J., Daneshvar, B., Lauridsen, S. T., Knuthsen, P., et al. (1999). Effect of fruit juice intake on urinary quercetin excretion and biomarkers of antioxidative status. *American Journal of Clinical Nutrition, 69*(1), 87–94.

Zhong, Y., Ma, C., & Shahidi, F. (2012). Antioxidant and antiviral activities of lipophilic epigallocatechin gallate (EGCG) derivatives. *Journal of Functional Foods, 4*(1), 87–93.

Zhong, Y., & Shahidi, F. (2011). Lipophilized epigallocatechin gallate (EGCG) derivatives as novel antioxidants. *Journal of Agricultural and Food Chemistry, 59*, 6526–6533.

Zhong, K., Shao, Z., & Hong, F. (2008). Enzymatic production of epigallocatechin by using an epigallocatechin gallate hydrolase induced from *Aspergillus oryzae*. *Biotechnology Progress, 24*(3), 583–587.

INDEX

Note: Page numbers followed by "*f*" indicate figures, and "*t*" indicate tables.